FOCUS ON PHYSICS

The Barnes & Noble Focus on Physics titles are prepared under the general editorship of J. WARREN BLAKER, Associate Professor of Physics, Vassar College.

ABOUT THE AUTHOR

Robert L. Stearns is a graduate of Wesleyan University and holds a Ph.D. from Case Institute of Technology. He has taught at Queens College and is currently Professor of Physics at Vassar College. In addition Professor Stearns holds an appointment as research collaborator with a nuclear physics group at Brookhaven National Laboratory.

College Outline Series

FOCUS ON PHYSICS
Mechanics I—
Statics, Dynamics, and Kinematics

ROBERT L. STEARNS

Professor of Physics
Vassar College

BARNES & NOBLE, INC., NEW YORK
Publishers   *Booksellers*   *Since 1873*

Printed in the United States of America

Preface

FOCUS ON PHYSICS

The ten Focus on Physics volumes in the College Outline Series are concise but comprehensive self-teaching treatments of the most important topics in the first-year physics course.

It has been found that beginning students of physics often require material which can supplement their texts and lectures by supplying, in a somewhat different format, explanations of principles and methods. Such re-enforcement is particularly important in areas where the student is having difficulty. A number of short "outline" books which emphasize the problem-solving aspect of physics are available for this purpose, but we know of no short works which provide a more thorough discussion of underlying principles.

Each Focus on Physics title has been planned to present the subject matter of a particular topic or group of related topics. Each book includes summaries of principles and facts, solved examples, problems with answers, and numerous illustrations.

In order to help the student to attain understanding and mastery of the basic material, these volumes strongly emphasize physical principles. They reflect the detailed approach to physics which is now the standard treatment, and their combined subject matter forms the essence of physical science.

In addition to use by the individual student, the Focus on Physics titles are suitable for class assignment as text supplements. They can also be used for self-instruction by the general reader who is interested in exploring and learning about the elements of modern physics.

J. WARREN BLAKER
General Editor

Table of Contents

Introduction

MECHANICS I—Statics, Dynamics, and Kinematics

Since mechanics is fundamental to all of physics it is important to introduce it in as simple and straightforward a way as possible. We have therefore used the standard approach of starting with the relatively simple—and hopefully familiar—concepts of kinematics, followed by an elementary treatment of vector analysis, which is then used in the discussion of more complicated problems such as projectile and circular motion. A full chapter is devoted to Newton's second law and its application. Although recent practice has eliminated discussion of statics from elementary physics textbooks in favor of other important topics, we think it appropriate to include statics in this volume which is entirely devoted to mechanics. The final three chapters, dealing with the slightly more sophisticated concepts of energy and momentum and the topics related to planetary motion, use many of the concepts introduced earlier.

Numerous examples with complete solutions are included throughout the text. Additional problems for each chapter to be worked out by students are included at the end of the book, accompanied by answers but not solutions.

Important topics such as wave motion, simple harmonic motion, and a more extended treatment of rotational motion will be found in Mechanics II of the Focus on Physics Series.

Chapter 1

Standards and Units

1-1. Introduction. The study of physics is usually divided into particular areas such as mechanics, electricity and magnetism, thermodynamics, optics, atomic physics, solid state physics and nuclear and particle physics. The systematic development of theories and techniques in all of these areas depends on the use of many fundamental, and probably familiar, concepts such as force, velocity, acceleration, energy, momentum and angular momentum. These concepts and the laws relating them might be said to constitute the field of mechanics. Mechanics is therefore the "language" of physics and in this sense is certainly the most fundamental of all the branches of physics.

We have mentioned some of the important concepts that must be used in any discussion of mechanics. Many of these, such as velocity, acceleration and force, are familiar terms from our everyday language. Often their meaning in physics is similar to that in everyday usage, but usually we must be much more precise. We will give the definitions for these terms in this chapter and those that follow, and the reader will find that all of the concepts or quantities we define can be categorized as either fundamental or derived quantities. The derived quantities can be defined or expressed completely in terms of the fundamental quantities. In our treatment of mechanics the fundamental quantities will be length, time, and mass; all other quantities can be expressed as some combination of these three. The precise specification or description of the fundamental quantities requires the use of standards of length and time, which are discussed in the following sections, and of mass, which is more completely discussed in Chapter 3 after the development of Newton's second law relating force, mass, and acceleration.

1-2. The Standard of Length. The technique for measuring a length is the rather obvious one of comparing this length or distance with some arbitrary and convenient standard of length. There have been many standards of length over the years of recorded history, and often they have been related to the size of some part of a man's body, such as his foot or arm. Standards,

1

however, must be both conveniently available and precise. In recent years the trend has been toward greater and greater precision. For many years the standard of length was the standard meter, originally taken to be one ten millionth of the distance from pole to equator along a great circle passing through Paris. A later, more precise definition specified the standard meter as the length of a bar of platinum-iridium alloy kept at the International Bureau of Weights and Measures near Paris, France. The distance between two marks on this bar, when at a temperature of 0° centigrade (the temperature at which water freezes), was defined as the standard meter. Secondary standards could be compared with this bar and distributed to users throughout the world. The foot, a unit of distance frequently used in engineering work in this country, is a secondary standard defined in terms of the meter. There are exactly three feet in the unit of distance called the yard, and one yard is defined as $\frac{3600}{3937}$ of the standard meter.

To permit greater precision as well as a more readily available standard, the standard meter was redefined in 1960 in terms of the wavelength (see Focus on Physics, *Optics II*, by J. W. Blaker) of the orange light emitted by the krypton atom of mass number 86. This wavelength can be measured to very great accuracy and the standard can be made available to all laboratories simply by observing the light from an electric discharge tube containing krypton 86. This new standard does not change the length of the standard meter by a measurable amount, since the wavelength of this line had been determined as accurately as possible in terms of the old standard meter. The standard meter is now equal to 1,650,763.73 wavelengths of the orange line in the ^{86}Kr spectrum. The centimeter is exactly $\frac{1}{100}$th of the standard meter.

1-3. The Standard of Time. A standard of time should be a cyclic phenomenon which repeats with great regularity. Of the many choices available in nature, the most obvious is the time required for the earth to rotate once on its axis. Until 1960, the standard of time was the mean or average solar day, this being the time between noon on one day, when the sun is at its zenith, and the corresponding time the next day. This was a rather undesirable standard, however, since the length of the day is not constant throughout the year, so that one must use the average length of the day. Furthermore, the average length of the day has been gradually increasing at the rate of about 0.001 seconds per rotation every 100 years. This might not seem like much of an effect, but over a period of several thousand years it results in cumulative effects amounting to hours. For this reason, in 1960, the year 1900 was chosen as the "standard" year for the definition of the so-called ephemeris second, which was a specified fraction (about 1/86,400) of the mean solar day of that year. Because of the difficulty of actually adjusting a clock in terms of this ephemeris second, the standard second was again changed in 1967 and defined in terms of the precisely measurable frequency of an atomic vibration. The standard second is now defined as the duration of 9,192,631,770 periods of the radiation resulting from a transition between two levels in the atom of cesium 133. As in the case of the standard meter, the new standard was chosen so as not to be measurably different, within our present capabilities, from the previous standard.

1-4. The Standard of Mass. Mass is a quantitative measure of inertia, which is the tendency of an object to resist changes in its motion. An object with a large mass is difficult to move and correspondingly difficult to stop. Since the weight of an object is proportional to its mass, a convenient way to measure the relative mass of two objects is to compare their weights. We discuss in detail the concepts of inertia, weight, and mass in Chapter 3, and merely note here that the standard of mass is the standard kilogram kept at the International Bureau of Weights and Measures near Paris. The kilogram has a weight of about 2.21 pounds.

Chapter 2
Kinematics

2-1. Introduction. Mechanics is concerned with the fundamental relationships between force, matter, and motion. In this chapter we will be concerned with that branch of mechanics known as kinematics, which is the description of motion independent of how the motion is produced. We begin with the concepts of speed and acceleration for motion limited to one direction. We then extend our discussion to more complicated situations, such as projectiles and circular motion, for which we must take into account the fact that velocity and acceleration are quantities that require the specification of both a magnitude (speed in the case of velocity), and a direction, if they are to be described completely. The relationships between force, mass, and acceleration will be discussed in Chapter 3.

2-2. Speed. The speed of a moving object is defined as the rate of change of distance with time experienced by that object. For example, if an automobile were to travel 10 feet in one second, we would say that its speed was 10 feet per second. If the automobile continues to travel 10 feet in every subsequent second, we say that its speed is constant and we can calculate how far it will travel in any length of time using the relationship:

$$d = vt. \tag{2.1}$$

Here we have used v (for velocity) to represent the speed, anticipating the fact that the speed is the magnitude of the velocity. Speed can be expressed in any convenient set of units. Some common units are miles per hour, meters per second, and feet per second.

Example 2-1. A truck traveling at constant speed requires 20 seconds to go 100 feet. What is the speed of the truck?

Solution. Using equation 2.1 we get $v = d/t = 100/20 = 5$ ft/sec.

Example 2-2. How far will the truck in Example 2-1 travel in one hour at the constant speed of 5 ft/sec?

Solution. We solve for the distance in equation 2.1 and get:

$$d = v \times t = 5 \text{ ft/sec} \times 60 \text{ sec/min} \times 60 \text{ min/hr} = 18,000 \text{ ft.}$$

Note that in order to use equation 2.1 correctly in Example 2-2, we must express the time in seconds to make the units consistent. When the velocity is expressed in ft/sec, the distance must be expressed in feet and the time in seconds.

Example 2-3. Convert the speed 60 miles per hour into units of feet per second.

Solution. We utilize the fact that in one mile there are 5280 ft, and that in one hour there are $60 \times 60 = 3600$ sec. Thus,

$$v = 60 \text{ mph} \times 5280 \frac{\text{ft}}{\text{mi}} \times \frac{1}{60} \frac{\text{min}}{\text{sec}} \times \frac{1}{60} \frac{\text{hr}}{\text{min}} = 88 \text{ ft/sec.}$$

This is a handy number to remember for converting between these two fairly common systems. For example, to express 50 mph in terms of ft/sec, we merely use a simple proportion, i.e.,

$$\frac{50}{60} = \frac{v}{88}, \text{ and } v = 88 \frac{\text{ft}}{\text{sec}} \times \frac{50 \text{ mph}}{60 \text{ mph}} = 73.3 \text{ ft/sec.}$$

2-3. Average Speed. In many situations the speed is not constant, and it is convenient to use an average speed to describe the rate at which distance is traveled as a function of time. The average speed is defined as the total distance traveled divided by the time required. The average speed is equal to that constant speed which would be required for an object to travel the same distance in the same time. We represent the average speed by $\bar{v}$ so that:

$$\bar{v} = \frac{d}{t}. \tag{2.2}$$

Example 2-4. At Easter vacation time a student leaves New York City for the sunny south. After exactly 24 hours of driving, including numerous stops, he finds that he has covered a total distance of 1000 miles. What was his average speed?

Solution. We use $\bar{v} = \dfrac{d}{t} = \dfrac{1000 \text{ mi}}{24 \text{ hr}} = 41.7$ mph, dividing the total distance by the total time.

Example 2-5. A car travels at the constant speed of 30 mph for 20 miles, speeds up to 40 mph for the next 20 miles, and then travels the final 20 miles at 50 mph. What was the average speed for the trip?

Solution. Here there is a temptation to jump to the conclusion that the average speed is 40 mph, but when we carefully apply our definition for average velocity, we find that this is not correct. The total distance here is clearly 60 miles but the total time is the sum of the times for each segment of the trip, i.e.,

$$t = \frac{20}{30} + \frac{20}{40} + \frac{20}{50} = 0.67 + 0.50 + 0.40 = 1.57 \text{ hrs.}$$

Therefore,

$$\bar{v} = \frac{d}{t} = \frac{60}{1.57} = 38.2 \text{ mph.}$$

Example 2-6. Calculate the average velocity for a trip which is made according to the following plan: 2 hours for the first 50 miles, 3 hours for the next 100 miles, 1 hour for the next 40 miles and 4 hours for the final 60 miles.

Solution. Here a straightforward application of the definition of average velocity yields:

$$\bar{v} = \frac{50 \text{ mi} + 100 \text{ mi} + 40 \text{ mi} + 60 \text{ mi}}{10 \text{ hr}} = \frac{250 \text{ mi}}{10 \text{ hr}} = 25 \text{ mph.}$$

2-4. Instantaneous Speed. Even though the speed of an object may be continually changing, it is often desirable, and in fact common practice, to specify what is called the instantaneous speed. This is the instantaneous rate of change of distance and is, for example, the quantity given by the speedometer on a car. Figure 2-1 is used to show exactly what we mean by instantaneous speed. We define instantaneous speed to be the value which the ratio $\Delta d/\Delta t$ approaches as Δt becomes very small and approaches zero.

$$v = \lim_{\Delta t \to 0} \frac{\Delta d}{\Delta t}. \tag{2.3}$$

Figure 2-1 is a graph of distance versus time for an object which moves between the points O and C. The dotted line between O and C has constant slope, representing constant speed, and, in fact, represents the average speed for the object which moved from O to C. However, if we want the instantaneous velocity at the point P, we note that as we calculate the ratio $\Delta d/\Delta t$ for the pairs of points AB, $A'B'$, $A''B''$, corresponding to smaller and smaller values of Δt, the ratio does not change much as Δt approaches zero. This ratio does not become infinite, as one might guess, but is simply the slope of the tangent to the curve at the point P.

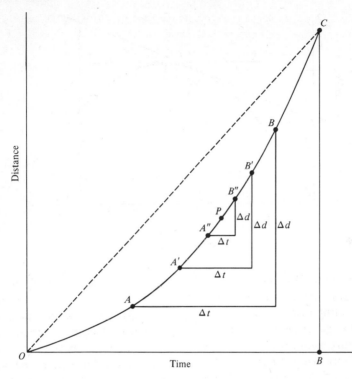

Figure 2-1.

Example 2-7. In Figure 2-2, the position of a moving object is plotted vs. the time. Estimate the instantaneous speed of this object at the point P.

Solution. The speed at the point P is given by the slope of the curve at that point. As indicated in Figure 2-2, the slope at this point is approximately $\frac{4}{9}$ meters per second.

Example 2-8. For the object moving as indicated in Figure 2-2, determine the time at which the speed is zero.

Solution. Since the speed is given at any point by the slope of the curve, the speed will be zero when the slope is zero, i.e., when the tangent to the curve is parallel to the time axis. As indicated on the figure, this occurs at a time of approximately 5.7 sec.

2-5. Acceleration. We are all familiar with the results produced by moderately rapid changes in the speed of an automobile

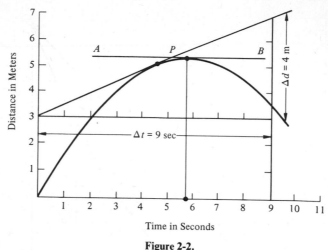

Figure 2-2.

or airplane. When such changes occur and we are thrown forward or backward in our seat, we say that the vehicle is accelerating. We define acceleration, for motion in a straight line, as the rate of change of velocity with time. Just as in the case of speed, we can talk about both an average and an instantaneous acceleration, but here we restrict our discussion to cases for which the acceleration is constant. Again we point out that although our discussion is limited to motion in a straight line—and for this case, the acceleration must be directed in the same sense or the opposite sense as the velocity—in general, the acceleration may have any direction, so that to specify it completely we must give both a magnitude and a direction. As a reminder that the speed is only the magnitude of the velocity, we continue to represent it with the letter v and obtain the relationship:

$$a = \frac{v}{t} \tag{2.4}$$

relating the speed, the acceleration and the time for an object which starts with speed zero when $t = 0$. A simple example should make the meaning of equation 2.4 clear. Suppose that a car, at rest at a traffic light, starts up in such a manner that its speed changes by 2 ft/sec every second. This means that its acceleration is 2 ft/sec^2 or two feet per second per second. Thus at the end of the first second its speed is 2 ft/sec, at the end of the

second second its speed is 4 ft/sec, increasing to 6, 8 and 10 ft/sec
etc. in successive one-second intervals. (*Cf.* Figure 2-3(b).)
   Equation 2.4 refers to the special case of constant acceleration.
It should be clear that it is quite possible for the speed to change
in a nonuniform manner as a function of time. In this case the
acceleration is not constant, and we can define an instantaneous
acceleration, $a = \lim\limits_{\Delta t \to 0} \dfrac{\Delta v}{\Delta t}$, just as we defined an instantaneous
velocity. Here, Δv is the velocity change occurring in the time Δt.
For constant acceleration this ratio, of course, does not change.
   Example 2-9.   A bicycle, starting from rest, rolls down a hill
at constant acceleration. It is observed that after 2 seconds, the
speed is 6 ft/sec. What is the acceleration and what will the speed
be after 10 seconds?
   Solution.   Using equation 2.4 we get:

$$a = \frac{v}{t} = \frac{6 \text{ ft/sec}}{2 \text{ sec}} = 3 \text{ ft/sec}^2.$$

If the acceleration is 3 ft/sec², the speed after 10 sec will be

$$v = at = 3 \text{ ft/sec}^2 \times 10 \text{ sec} = 30 \text{ ft/sec}.$$

2-6. Rectilinear Motion with Constant Acceleration.   Thus far
in our discussion of kinematics we have encountered the variables
speed, acceleration, time and distance. These quantities are not
independent and it will be instructive for us to investigate the
relationships among them for the special case of motion in a
straight line with constant acceleration. To be somewhat more
general than we have been up to this point, we will want to
include a fifth variable, the initial velocity, which we have, until
now, taken to be zero.
   In this section we shall obtain an expression for the speed in
terms of the initial velocity, the acceleration, and the time; the
distance in terms of the initial velocity, the acceleration and the
time; and, finally, the velocity in terms of the initial velocity,
the acceleration, and the distance. First, however, it will be useful
to investigate the significance of the area under a graph of speed
vs. time. It turns out that this area is equal to the distance traveled
by the object and we show this using Figure 2-3(a). Consider the
rectangle of width Δt. The area of this rectangle is $v\Delta t$ and this
value is very close to the distance traveled by the object in the
time Δt. As Δt gets smaller and smaller, this area becomes a

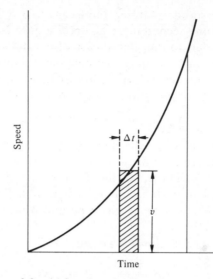

Figure 2-3.   (a) Speed vs. time for variable acceleration.

better and better approximation to the distance traveled. We can imagine that the area under the curve is made up of many, many such rectangles of width Δt. The sum of the areas of these rectangles approaches the area under the curve as we make Δt for each very small. Thus, it should be clear that as Δt approaches the limiting value of zero, the sum of the areas of the rectangles approaches the area under the curve, and this must be the distance traveled from the beginning to the end of the time interval for which this area is calculated.

Figure 2-3(b) illustrates another interesting point for the special case of constant acceleration, namely, that the average velocity for any time interval must be midway between the initial and final velocity for that interval. This follows, of course, from the fact that the speed varies linearly with time so that it is perfectly proper to calculate the average velocity by adding the initial velocity and the final velocity and dividing by two. Figure 2-3(b) shows this graphically. The average speed v is half the final speed and the area of the rectangle $ODCE$ is the same as the area of the triangle OBE, indicating that the distance traveled by the object is the same as if the speed had been fixed at the constant value

$$\bar{v} = \frac{v_i + v_f}{2} = \frac{O + v_f}{2} = \frac{v_f}{2} \text{ for an initial velocity of zero.}$$

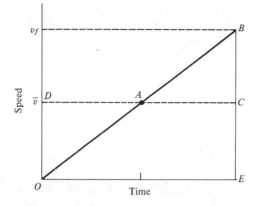

Figure 2-3. (b) Speed vs. time for constant acceleration.

Example 2-10. A racing car passes one end of the grandstand at a speed of 50 ft/sec and slows down at constant acceleration to a speed of 10 ft/sec as it passes the other end of the grandstand. If this process takes 20 seconds, calculate the acceleration and determine graphically how long the grandstand is.

Solution. Since $a = \Delta v/\Delta t$ we get $a = -40$ ft/sec/20 sec = -2 ft/sec^2. Since the average velocity must be (50 ft + 10 ft)/2 sec = 30 ft/sec, and the grandstand length must be given by $d = \bar{v}t$ (using equation 2.2), we get $d = 30$ ft/sec × 20 sec = 600 ft. Figure 2-4 illustrates how this result is obtained graphically. The area under the curve is the area of triangle ABC plus the area of the rectangle $BCDO$. These areas are: $\frac{1}{2}$ × 40 ft/sec × 20 sec = 400 ft, and 10 ft/sec × 20 sec = 200 ft, their sum being 600 ft, which agrees with the distance obtained above.

Next, let us obtain an expression which is a more general version of equation 2.4. We wish to treat the situation in which the initial speed is not zero but has some arbitrary value, v_0. It should be apparent, by inspection, that we need only modify equation 2.4 by adding the initial speed such that:

$$v = v_0 + at \qquad (2.5)$$

since the change in the speed in a time t is given by the expression at to which we merely add v_0. It is important to note, however, that the sign between the two terms in equation 2.5 may be either positive, as indicated, or negative, depending upon whether the

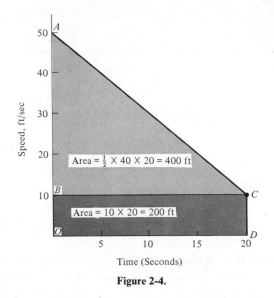

Figure 2-4.

acceleration is in the same sense or is oppositely directed relative to the initial velocity.

Example 2-11.   A ball rolls up an inclined plane with an initial velocity of 90 cm/sec. (There are 100 cm in 1 meter). It slows down at the rate of 30 cm/sec². Calculate its velocity at the end of 2, 3, and 6 seconds.

Solution.   Clearly, the initial speed and the acceleration are oppositely directed so that we must use a minus sign in equation 2.5 if we arbitrarily denote the direction up the plane as positive. For the three times specified we get:

$$t = 2 \text{ sec}, v = 90 \text{ cm/sec} - 30 \text{ cm/sec}^2 \times 2 \text{ sec} = 30 \text{ cm/sec}$$
$$t = 3 \text{ sec}, v = 90 \text{ cm/sec} - 30 \text{ cm/sec}^2 \times 3 \text{ sec} = 0$$
$$t = 6 \text{ sec}, v = 90 \text{ cm/sec} - 30 \text{ cm/sec}^2 \times 6 \text{ sec} = -90 \text{ cm/sec}$$

indicating that after 3 seconds the ball stops and rolls back down the plane and has a velocity of 90 cm/sec in the direction opposite to its original motion after 6 seconds. It becomes clear that the distance traveled up the plane during the first 3 seconds is equal to the distance traveled down the plane in the second 3 second interval because the average velocity ($v_i + v_f/2$) is the same (45 cm/sec) in each case. The ball will therefore be back at its starting point at the end of 6 seconds.

Next we obtain an expression for the distance traveled in terms of the acceleration, the time and the initial speed, v_0. To do this we utilize the method for calculating average speed mentioned above for the case of constant acceleration and obtain:

$$d = \bar{v}t = \frac{(v_0 + v)}{2} \times t$$

in which we have replaced $\bar{v}$ with $\frac{(v_0 + v)}{2}$. Since we know that $v = v_0 + at$ we can write:

$$d = \frac{(v_0 + v_0 + at)t}{2}$$

$$d = v_0t + \tfrac{1}{2}at^2. \tag{2.6}$$

Again, it is important to realize that v_0 and a may be oppositely directed, i.e., the acceleration may result in either an increase or decrease in the speed so that the sign between the two terms in equation 2.6 may be either positive, as shown, or negative. It is also important to understand that equation 2.6 gives the *displacement from the starting* point and not the total distance traveled. Example 2-12 illustrates this point.

Example 2-12. Consider the situation described in Example 2-11 in which a ball rolls up an inclined plane. Calculate the displacement from the starting point at the end of 2, 3, and 6 seconds.

Solution. Using equation 2.6 and the values $a = -30$ cm/sec^2 and $v_0 = 90$ cm/sec, we get for the three times:

For 2 seconds:

$$s = v_0t + \tfrac{1}{2}at^2 = 90 \times 2 - \tfrac{1}{2} \times 30 \times (2)^2 = 120 \text{ cm}.$$

For 3 seconds:

$$s = v_0t + \tfrac{1}{2}at^2 = 90 \times 3 - \tfrac{1}{2} \times 30 \times (3)^2 = 135 \text{ cm}.$$

For 6 seconds:

$$s = v_0t + \tfrac{1}{2}at^2 = 90 \times 6 - \tfrac{1}{2} \times 30 \times (6)^2 = 0 \text{ cm}.$$

The total distance traveled must be twice 135 cm or 270 cm since the ball comes to rest, as shown in Example 2-11, after 3 seconds and rolls back down to its original position after another 3 seconds. Thus, care must be taken to distinguish between the total distance traveled in either direction and the net displacement from the starting point, which is the quantity given by equation 2.6.

Example 2-13. A driver traveling at 60 miles per hour suddenly sees a dog at a distance of 180 ft. If he applies the brakes immediately and it is known that the maximum possible acceleration is 22 ft/sec^2, will he be able to stop in time?

Solution. We use equation 2.6 after first calculating the time required. Since 60 mph is equal to 88 ft/sec and $a = \Delta v/\Delta t$, we get

$$t = \Delta t = \frac{\Delta v}{a} = \frac{88 \text{ ft/sec}}{22 \text{ ft/sec}^2} = 4 \text{ sec.}$$

Thus,

$$d = v_0 t + \tfrac{1}{2}at^2 = 88 \text{ ft/sec} \times 4 \text{ sec} - \tfrac{1}{2} \times 22 \text{ ft/sec}^2 \times (4)^2 \text{ sec}^2$$

$$d = 352 \text{ ft} - 176 \text{ ft} = 176 \text{ ft}$$

showing that the dog is safe. Again, note that in this case v_0 and a were oppositely directed. Note that the problem is also easily solved using the fact that the average velocity during the four seconds used to stop must have been $(v_0 + 0)/2 = 88/2 = 44$ ft/sec. Therefore,

$$d = \bar{v}t = 44 \times 4 = 176 \text{ ft.}$$

This procedure, of course, was the basis for the derivation of equation 2.6.

Example 2-14. A boy standing on a bridge over a river 104 feet below throws a stone straight down and observes that it takes 2 seconds to hit the river. Assuming that the acceleration experienced by the stone is 32 ft/sec^2, calculate the initial velocity.

Solution. We note that in this case the acceleration and initial velocity are in the same sense and apply equation 2.6 directly.

$$d = v_0 t + \frac{1}{2}at^2 \text{ and } v_0 = \frac{d}{t} - \frac{1}{2}at$$

$$v_0 = \frac{104 \text{ ft}}{2 \text{ sec}} - \frac{1}{2} \times 32 \text{ ft/sec}^2 \times 2 \text{ sec} = 52 \text{ ft/sec} - 32 \text{ ft/sec}$$

$$v_0 = 20 \text{ ft/sec.}$$

It is also possible to obtain an expression in terms of velocity, acceleration and distance by again using the expression

$$d = \bar{v}t = \frac{(v_0 + v)}{2} t$$

as we did in the derivation of equation 2.6. Here we use the relationship $v - v_0 = at$ but eliminate t instead of v. Since

$$t = \frac{(v - v_0)}{a},$$

we get:

$$d = \left(\frac{v + v_0}{2}\right)\left(\frac{v - v_0}{a}\right) = \frac{v^2 - v_0^2}{2a}$$

and

$$v^2 - v_0^2 + 2ad. \tag{2.7}$$

Example 2-15. A motorcycle can accelerate at the rate of 30 ft/sec². It is desired to cover a distance of 100 ft such that the speed at the end of this distance is 80 ft/sec. What initial velocity is required?

Solution. From equation 2.7 we obtain $v_0^2 = v^2 - 2ad$,

$$v_0^2 = (80)^2 - 2 \times 30 \text{ ft/sec}^2 \times 100 \text{ ft} = 6400 - 6000$$

$$= 400 \text{ ft}^2/\text{sec}^2$$

$$v_0 = 20 \text{ ft/sec.}$$

Example 2-16. Two motorcycles are at rest and separated by 24.5 ft. They start at the same time in the same direction, one having an acceleration of 3 ft/sec², the other having an acceleration of 2 ft/sec², with the more rapidly accelerating machine overtaking the other. How long does it take for the faster cycle to overtake the slower, how far does the faster machine go before it catches up, and how fast is each cycle going at this time?

Solution. Each cycle travels for the same length of time but the overtaking machine must go an extra 24.5 ft. Therefore, we equate the distance traveled by the faster machine with the distance traveled by the slower plus 24.5 ft.

$$\tfrac{1}{2} a_1 t^2 = \tfrac{1}{2} a_2 t^2 + 24.5 \text{ ft}$$

$$\tfrac{1}{2} \times 3 \text{ ft/sec}^2 \times t^2 = \tfrac{1}{2} \times 2 \text{ ft/sec}^2 \times t^2 + 24.5 \text{ ft}$$

$$t^2 = 49 \text{ sec}^2 \text{ and } t = 7 \text{ seconds.}$$

The faster machine goes a distance given by:

$$d = \tfrac{1}{2} a_1 t^2 = \tfrac{1}{2} \times 3 \text{ ft/sec}^2 \times (7)^2 \sec^2 = 73.5 \text{ ft.}$$

The velocity of each after 7 seconds is obtained by using equation 2.4, i.e., $v = at$.

For $a = 3 \text{ ft/sec}^2 \qquad v = 3 \text{ ft/sec}^2 \times 7 \sec = 21 \text{ ft/sec.}$

For $a = 2 \text{ ft/sec}^2 \qquad v = 2 \text{ ft/sec}^2 \times 7 \sec = 14 \text{ ft/sec.}$

2-7. Free Fall.   Perhaps the most common example of motion in a straight line with constant acceleration is that of an object which falls freely due to the force of gravity. The law of universal gravitation will be discussed extensively in Chapter 8.   Here we simply present the experimental fact that all objects near the surface fall toward the center of the earth with an acceleration which varies slightly from place to place on the earth but always has a value very close to 9.8 m/sec^2 or 32 ft/sec^2. These are the values we will adopt for computations involving gravitational acceleration, which is usually designated by the letter g.

Example 2-17.   With what speed must a ball be thrown directly upward so that it remains in the air for 10 seconds? How high will it go and what will its speed be when it hits the ground?

Solution.   The problem is perhaps most easily approached by thinking about what must happen to the ball and then using equation 2.5. When the ball is thrown, its speed must decrease by 32 ft/sec each second until it reaches its maximum height. Then it starts to fall, gaining speed at the rate of 32 ft/sec^2 and retraces its path, hitting the ground with the same speed at which it started the trip upward. This is so because the acceleration is constant and the distance traveled is the same during the rising and falling portions of the motion of the ball. Thus, the average velocity must have the same magnitude in each case and the time required to reach the maximum height must equal the time to fall back to the ground. Thus, in equation 2.5, if we designate the upward direction as positive, v_0 will be positive and a negative. After 10 seconds, v must equal $-v_0$ and we get:

$$-v_0 = v_0 - at \quad \text{or} \quad -2v_0 = -10 \sec \times 32 \text{ ft/sec}^2$$

so that

$$v_0 = \frac{(10 \times 32)}{2} = 160 \text{ ft/sec.}$$

The height reached by the ball is easily obtained by realizing that the upward portion takes 5 seconds and the average velocity for this part of the motion must be $\bar{v} = (160$ ft/sec $+ 0)/2 = 80$ ft/sec. Therefore, the height is equal to $\bar{v}t = 80$ ft/sec $\times$ 5 sec $= 400$ ft. The same result is obtained by considering the downward trip and using $d = \frac{1}{2}at^2 = \frac{1}{2} \times 32$ ft/sec$^2 \times (5$ sec$)^2 = 400$ ft.

Example 2-18. A ball is dropped from a bridge 122.5 meters above a river. After the ball has been falling for two seconds, a second ball is thrown straight down after it. What must its initial velocity be so that both hit the water at the same time and what is the velocity of each at this time?

Solution. The first ball requires a time t to reach the water, where t is given by $d = \frac{1}{2}at^2$ or

$$t = \sqrt{\frac{2d}{a}} = \sqrt{\frac{(2 \times 122.5)}{9.8}} = \sqrt{25} = 5 \text{ sec.}$$

Therefore, for the second ball, only three seconds are available to travel the same distance, and using equation 2.6 we get:

$$d = v_0 t + \frac{1}{2}at^2 = 122.5 = 3v_0 + \frac{1}{2} \times 9.8 \times (3)^2$$

$$3v_0 = 122.5 \text{ m/sec} - 44.1 \text{ m/sec} = 78.4 \text{ m/sec}$$

$$v_0 = 26.1 \text{ m/sec.}$$

The velocity of the first ball when it hits the water is given by equation 2.4 and is $v = at = 9.8$ m/sec$^2 \times 5$ sec $= 49$ m/sec. The velocity of the second ball is given by equation 2.5, which gives $v = v_0 + at = 26.1$ m/sec $+ 9.8$ m/sec$^2 \times 3$ sec $= 26.1$ m/sec $+ 29.4$ m/sec $= 55.5$ m/sec.

Example 2-19. With what speed must an object be thrown at right angles to the surface of the earth (i.e., straight up) so that its maximum height above the earth is 144 ft?

Solution. We use equation 2.7 and call the upward direction positive. When the object reaches the maximum height its velocity is zero. Note also that the acceleration is negative, being directed toward the center of the earth. Equation 2.7 gives $v^2 = v_0^2 + 2as$ and with the above values becomes $0 = v_0^2 - 2gd$. With $g = 32$ ft/sec^2 and $d = 144$ ft we get: $v_0^2 = 2 \times 32 \times 144$ or $v_0 = \sqrt{(64 \times 144)} = 8 \times 12 = 96$ ft/sec.

Chapter 3
Vectors, Projectiles and Circular Motion

3-1. Vectors and Scalars. Our discussion of kinematics up to this point has been concerned only with motion in a straight line and we have avoided the problems which arise when the direction of such quantities as velocity and acceleration change. Velocity and acceleration are called vector quantities because both their magnitude and direction must be specified in order to give a complete description of them.* Force and momentum, which are discussed in subsequent chapters, are also vectors. Such concepts as time, energy, and mass are called scalars because there is no direction associated with them and it is possible to describe them using only their magnitude. In this chapter we will develop rules for the combination of vector quantities which will permit us to add and subtract forces or velocities, for example, and properly take account of both their magnitude and direction. We then apply some of what we have learned to describe projectile motion and circular motion, two common types of non-linear motion.

3-2. Displacement—an Example of a Vector. To acquire a feeling for the manner in which vector quantities combine, we use the displacement vector as a convenient example. Figure 3-1 illustrates what we mean by a displacement vector. We might imagine that a football player, starting at the point O at one corner of the field, walked out onto the field along the dotted line and arrived at the point P. The displacement vector **D,** serves to locate that point with respect to the sides of the field which serve as the x and y axes. Using the angle Θ and the length of the straight line OP, one can uniquely specify the location of point P or, by changing these two variables, the location of any point on the field. The displacement vector says nothing about the route taken by the football player in arriving at point P, but merely locates that point. Clearly, both a magnitude and a direction are required.

---

*In this and following chapters we adopt the convention of indicating vector quantities by boldface type. When boldface is not used in passages or equations concerned with vectors we are dealing with *the magnitude only* of a vector or with scalar quantities.

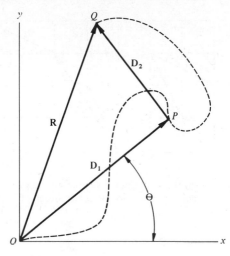

Figure 3-1. The displacement vector.

Suppose the football player were to continue his journey about the field and that we wished to locate other points along his path, such as point Q. The displacement of Q, relative to P is given by the vector $\mathbf{D}_2$. Note that by arranging the vectors $\mathbf{D}_1$ and $\mathbf{D}_2$ "head to tail" and connecting the tail of $\mathbf{D}_2$ (point P) to the head of $\mathbf{D}_1$, as shown in Figure 3-1, we obtain the resultant displacement, $\mathbf{R}$, of point Q from the origin (O). Thus, we have found one technique for combining two vectors so as to obtain a third or resultant which produces the same result (displacement in this case) as the original two taken together. This process is called vector addition. Figure 3-2(a) illustrates the addition or combination of four displacement vectors, using this "head to tail" method. Figure 3-2(b) shows that the order in which the vectors are combined or added makes no difference.

The "head to tail" method is perhaps not quite as useful as the method which makes use of the rectangular components of the vectors. This technique is illustrated in Figure 3-3 where a velocity vector is used. The length of the arrow represents the magnitude of the velocity, i.e., its speed, while the angle Θ gives its direction, relative to the x axis. The x component of this vector is the projection of v on the x axis and is given by $v_x = v \cos \Theta$. Similarly, the y component of the velocity is $v_y = v \sin \Theta$. It is apparent that the "head to tail" method gives v as the resultant when v_x is added to v_y. The use of these components along the

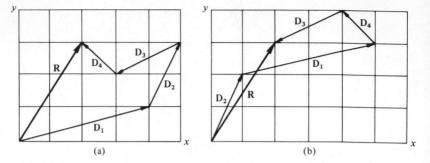

Figure 3-2.   In (a) and (b) the same four displacement vectors are added, but in different order, so as to show that the order of addition makes no difference.

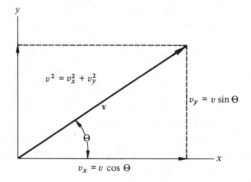

Figure 3-3.   The relationship between a vector and its rectangular components.

coordinate axes becomes very convenient when two or more vectors are combined because each vector can be broken up into its x and y components, and x components and y components can be added separately and algebraically. One thus obtains the x component of the resultant $\mathbf{R}_x$ (the algebraic sum of the x components), and the y component of the resultant $\mathbf{R}_y$ (the algebraic sum of the y components), and these can be combined to obtain the magnitude $\mathbf{R}$ of the resultant, using the Pythagorean theorem, $R = R_x^2 + R_y^2$, as well as the direction of the resultant, using $\tan \Theta = R_y/R_x$.

Example 3-1.   Consider the four displacement vectors shown in Figure 3-4. Use the method of components to obtain the vector sum. Give both the magnitude and direction of the resultant vector.

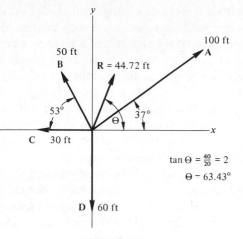

Figure 3-4.

Solution. We construct a table (3.1) as shown in which we list the x and y components of each vector. Vector **A** has an x component given by $100 \times \cos 37° = 100 \times 0.8 = 80$ ft. The y component of vector **A** is $100 \times \sin 37° = 100 \times 0.6 = 60$ ft. The sin

TABLE 3-1.

Vector	x comp.	y comp.
A	80 ft	60 ft
B	−30 ft	40 ft
C	−30 ft	0
D	0	−60 ft
Sum	+20 ft	+40 ft

and cos of 53° are 0.8 and 0.6, respectively, so that the x and y components of vector **B** are −30 ft and +40 ft, respectively. Vectors **C** and **D** are along one of the coordinate axes and thus have components along only one axis. When the x and y components are added algebraically, the x component is +20 and the y component +40, giving a resultant of $\sqrt{20^2 + 40^2} = 44.7$ ft. The tangent of Θ, the angle between **R** and the x axis is $\frac{40}{20} = 2$, giving $\Theta = 63.4°$.

3-3. Relative Velocity. An interesting application of the techniques for combining vectors arises in problems dealing with relative velocity. Problems in this category include that of an airplane flying above the earth at some angle with respect to the wind, or the problem of a boat moving across a river in which the current must be considered. The technique that is applied, in the case of the airplane, for example, is to say that the velocity of the plane with respect to the earth is equal to the velocity of the plane with respect to the air, plus the velocity of the air with respect to the earth. Each of these velocities, of course, is a vector, and therefore must be added using the "head to tail" or component method discussed in the preceding section. A rather trivial example of this is provided by examining the motion of an airplane flying north at 200 mph with respect to the air into a 50 mph wind blowing from north to south. Clearly, the two velocity vectors are oppositely directed, and when combined yield a velocity of 150 mph for the plane with respect to the earth. Equation 3.1,

$$\overrightarrow{v_{AC}} = \overrightarrow{v_{AB}} + \overrightarrow{v_{BC}} \tag{3.1}$$

represents the more general situation, indicating that the velocity of object A with respect to object C is given by the velocity of A with respect to object B, plus the velocity of B with respect to C, the arrows above the velocities indicating that this addition must be vector addition.

The following problems illustrate the application of this technique to situations which are somewhat more complicated.

Example 3-2. An airplane travels, with respect to the air, with a velocity of 150 mph. The wind blows from the northwest at 50 mph. In what direction must the pilot head in order to fly due north with respect to the earth? What is his speed with respect to the earth?

Solution. The vectors must combine as shown in Figure 3-5. One approach is to note that the easterly component of the wind, which is 50 cos 45° = 50 × 0.707 = 35.35 mph, must equal in magnitude the westerly component of the velocity of the plane with respect to the air, i.e.,

$$150 \sin \Theta = 35.35 \text{ mph}$$

$$\sin \Theta = \frac{35.35}{150} = 0.2357$$

$$\text{and } \Theta = 13.63°.$$

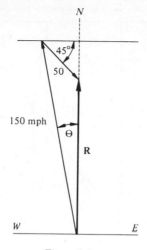

Figure 3-5.

The speed with respect to the earth is given by the magnitude of the vector **R** which must be:

$$R = 150 \cos \theta - 50 \sin 45° = 150 \times 0.9718 - 50 \times 0.707$$
$$R = 145.8 - 35.35 = 110.4 \text{ mph.}$$

Example 3-3. A boat travels at 4 ft/sec with respect to the water. This boat heads directly across a river in which the current is uniform and has the value of 3 ft/sec. The river is 80 ft wide at all points (Figure 3-6(a)).

(a) How long does the boat take to cross the river?

(b) What is the magnitude and direction of the velocity of the boat with respect to the banks as it crosses the river?

(c) How far downstream, from a point directly across from the starting point, does the boat reach the far shore?

(d) At what angle, with respect to a line perpendicular to the banks, would the boat have to head in order to move directly across the river?

Solution. (a) Although the boat moves downstream with the current at 3 ft/sec the time to cross is determined by the velocity component perpendicular to the banks. This is 4 ft/sec and thus the time is $t = \frac{80}{4} = 20$ sec.

(b) The boat has a velocity component of 4 ft/sec at right

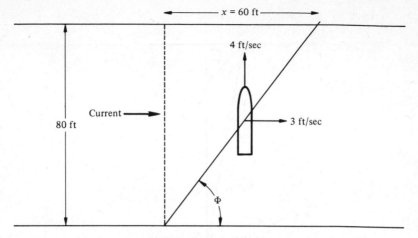

(a)

Example 3-3, parts (a), (b), (c).

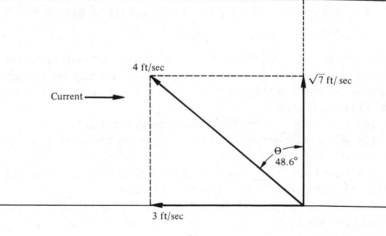

(b)

Example 3-3, part (d).

Figure 3-6.

angles to the banks and a component of 3 ft/sec parallel to the banks. Thus the magnitude of the velocity is $v = \sqrt{(3)^2 + (4)^2} = 5$ ft/sec. The angle Φ with respect to the banks has a tangent of $\frac{4}{3}$ and is thus 53°.

(c) Since the trip across the river requires 20 sec, the distance traveled downstream must be $x = v_x t = 3 \times 20 = 60$ ft, because the velocity component parallel to the banks is 3 ft/sec.

(d) In order to move directly across the river, i.e., at right angles to the bank, the boat must head upstream at an angle Θ such that the component of the velocity of the boat with respect to the water parallel to the banks is equal and opposite to the current. If this is to be so, the magnitude of this component must be 3 ft/sec. Figure 3-6(b) indicates that the sin of the angle Θ must be $\frac{3}{4}$ so that Θ is 48.6°. Figure 3-6(b) also indicates that the magnitude of the velocity component of the boat at right angles to the bank must be $\sqrt{7}$ ft/sec. This is obtained by applying the Pythagorean theorem to the vector triangle in Figure 3-6(b).

Example 3-4. A boat travels directly upstream in a river, moving at constant but unknown speed with respect to the water. At the start of this trip upstream, a bottle is dropped over the side. After 15 minutes the boat turns around and heads downstream. It catches up with the bottle when the bottle has drifted one mile downstream from the point at which it was dropped into the water. What is the current in the stream?

Solution. This problem becomes considerably easier if approached from the point of view of a coordinate system at rest with respect to the water in the river. In such a system the water is at rest and the banks appear to move upstream. The bottle is at rest with respect to the water and thus it should be clear that the return trip downstream must also take 15 minutes, since from the point of view of this coordinate system, it is just as though the boat were moving in a perfectly still pond. Once it is known that the round trip takes half an hour, it is obvious that the current in the river must be 2 miles per hour since the bottle moves one mile in a half hour.

3-4. Projectile Motion. The motion of an object such as a baseball or golf ball which is thrown or hit into the air above the surface of the earth is easily treated in terms of the concepts which have been discussed up to this point. We idealize our treatment and neglect the effect of air resistance which can have an extremely important effect on this type of motion if the velocities involved are fairly large. The results of this treatment will be a fairly good approximation for relatively heavy objects moving with sufficiently small velocities, such that the forces due to air resistance are small compared to the weight of the projectile. Our approach will be to consider an object projected in the air with some initial velocity v_0 directed at an angle Θ with respect to a horizontal earth. We can consider the vertical and horizontal portions of the motion to be independent in our analysis, being connected only by the fact that the object is in the air for a fixed time common to both motions. Perhaps the easiest way to show that the horizontal and vertical components of projectile motion are independent is to imagine a situation such as the following. Consider a flatcar moving by a railroad platform and a man sitting in a chair on the moving flatcar. The man throws a ball directly upward. It goes straight up and comes back down into his hand. There is no horizontal component with respect to the car. When the ball is observed by someone on the platform, however, it appears to move as a projectile and a constant horizontal motion is superimposed on the vertical motion. This vertical motion, however, is unchanged, and the maximum height to which the ball rises and the vertical velocity at any point appear the same to the man on the platform as to the man on the moving car. Thus, the presence of a horizontal component of the velocity has no effect on the vertical component, and we should be able to analyze them separately.

As indicated in Figure 3-7, we break up the initial velocity into x and y components, $v_0 \cos \Theta$, and $v_0 \sin \Theta$, respectively. We use these components to treat the horizontal and vertical motions separately, starting with the vertical motion. Let us obtain answers to the following questions: How long is the projectile in the air and what is the maximum height h that it attains? Considering only the vertical motion, the particle moves up, slowing down at the rate of 9.8 m/sec^2 or 32 ft/sec^2 until its velocity is zero at height h. It then starts to fall with the same acceleration and strikes the ground with a velocity equal and opposite to v_y. From the symmetry of the motion it should be clear that the time required to reach the height h is the same as that required to fall back to the ground. The time to reach the height h is simply given by $t_h = v_y/g$, and thus the time spent in the

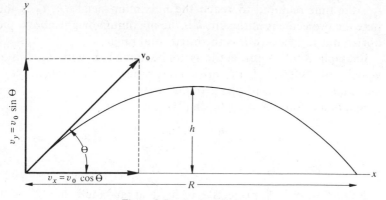

Figure 3-7. Projectile motion.

air is $2v_y/g$. Knowing the time required to reach the height h, it is easy to calculate the height h. We use $h = \bar{v}_y t_h$ where $\bar{v}_y$ is the average value of the y component of the velocity. The y component changes linearly from v_y to zero and therefore,

$$\bar{v}_y = \frac{(v_y + 0)}{2} = \frac{v_y}{2}.$$

Thus,

$$h = \bar{v}_y t_h = \left(\frac{v_y}{2}\right)\left(\frac{v_y}{g}\right) = \frac{v_y^2}{2g}.$$

We are now in a position to calculate the range R. The x component of the velocity, $v_x = v_0 \cos \Theta$, does not change and we can write $R = v_x(2t_h) = v_x \times 2v_y/g$, or $R = (2v_0^2 \sin \Theta \cos \Theta)/g = (v_0^2/g) \sin (2\Theta)$ since $2 \sin \Theta \cos \Theta = \sin (2\Theta)$.

Since the maximum value of $\sin (2\Theta)$ occurs for $\Theta = 45°$, the maximum range for any v_0 occurs for $\Theta = 45°$. Table 3-2 summarizes the results obtained above for a projectile launched at an angle Θ above a horizontal surface.

TABLE 3-2.

t_h	t_R	h	R	
$\dfrac{v_y}{g}$	$\dfrac{2v_y}{g}$	$\dfrac{v_y^2}{2g}$	$\dfrac{2v_0^2}{g} \sin \Theta \cos \Theta$	
			$\dfrac{v_0^2}{g} \sin (2\Theta)$	

t_h is the time required to reach the maximum height h; t_R is the time the projectile is in the air; h is the maximum height above the horizontal surface and R is the horizontal range.

Example 3-5. A projectile is to be launched with an initial velocity of 80 ft/sec. If the horizontal range is to be 160 ft, what angle must v_0 make with the ground?

Solution. Since $R = v_0^2/g \sin(2\Theta)$ we obtain:

$$\sin(2\Theta) = \frac{Rg}{v_0^2} = \frac{(160 \times 32)}{(80)^2} = 0.8$$

$$2\Theta = 53.1°, \quad \text{and} \quad \Theta = 26.6°.$$

Example 3-6. A projectile is fired at an angle of 37° with respect to the horizontal with an initial velocity of 50 m/sec. How long is it in the air, how high does it go and what is the horizontal range?

Solution. The length of time in the air, from Table 3-2, is given by:

$$t_R = \frac{2v_y}{g} = \frac{(2v_0 \sin \Theta)}{g}$$

$$t_R = \frac{(2 \times 50 \times 0.6)}{9.8} = 6.12 \text{ sec.}$$

The height h is given by:

$$h = \frac{v_y^2}{(2g)} = \frac{v_0^2 \sin^2 \Theta}{(2g)}$$

$$h = (50)^2 \times \frac{(0.6)^2}{(2 \times 9.8)}$$

$$h = 45.9 \text{ ft}$$

The horizontal range $R = \dfrac{2v_0^2 \sin \Theta \cos \Theta}{g}$

$$R = \frac{2 \times (50)^2 \times 0.6 \times 0.8}{9.8} = 245 \text{ m.}$$

Example 3-7. A projectile is shot at an angle of 30° with the horizontal, with an initial velocity of 100 ft/sec as shown in Figure 3-8.

If it is launched 40 ft from the top of the cliff, which is 30 ft high, how far does it land from the base of the cliff and how long was it in the air?

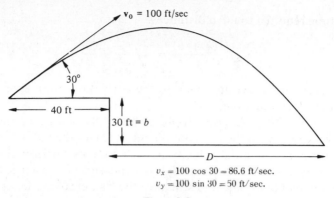

$$v_x = 100 \cos 30 = 86.6 \text{ ft/sec.}$$
$$v_y = 100 \sin 30 = 50 \text{ ft/sec.}$$

Figure 3-8.

Solution. Perhaps the easiest approach is to find first the time in the air, then find D using $D = v_x t - 40$ ft. Equation 2.6 gives:

$$-b = v_y t - \tfrac{1}{2} g t^2$$

which gives, after slight rearrangement:

$$t^2 - \frac{2 v_y t}{g} - \frac{2b}{g} = 0.$$

Using the general solution for a quadratic equation we get:

$$t = \frac{\dfrac{2 v_y}{g} \pm \sqrt{\dfrac{4 v_y^2}{g^2} + \dfrac{8b}{g}}}{2} = \frac{v_y}{g} + \sqrt{\frac{v_y^2}{g^2} + \frac{2b}{g}}.$$

Since the minus sign before the radical yields a negative value for t, which has no physical significance, we use the plus sign and get:

$$t = \frac{50}{32} + \sqrt{\frac{(50)^2}{(32)^2} + \frac{2 \times 30}{32}} = 1.56 + 2.08 = 3.64 \text{ sec}$$

$$R = v_x t = v_0 \cos 30° \times 3.64 \text{ sec} = 86.6 \times 3.64 = 315.2 \text{ ft}$$

$$D = R - 40 = 315 - 40 = 275 \text{ ft.}$$

Note that the terms $\dfrac{v_x v_y}{g}$ and $v_x \sqrt{\dfrac{v_y^2}{g^2} + \dfrac{2b}{g}}$ are the distances that the projectile travels horizontally before and after reaching its highest point, respectively. The time to reach the highest point is v_y/g and the height h above the top of the cliff is $v_y^2/2g$, so that the time required to fall a distance $h + b$ is

$$\sqrt{\frac{2(h + b)}{g}} = \sqrt{\frac{v_y^2}{g^2} + \frac{2b}{g}}.$$

This is an alternative approach to the derivation of an expression for the horizontal range.

Example 3-8. A baseball player hits a fly ball at an angle of 45° with respect to the horizontal field. It leaves the bat 4 ft above the ground. The initial velocity is such that its horizontal range $(v_0^2/g) \sin 2\theta$ is 400 ft. How high is it possible for a fence to be 360 ft from home plate and still have the ball clear the fence for a home run?

Solution. Here $\sin (2\theta) = 1$ so that $400 = v_0^2/g$ and $v_0 = \sqrt{12,800}$. $\sin 45° = \cos 45° = 1/\sqrt{2}$, so that $v_x = v_0 \cos 45° = v_y = v_0 \sin 45° = 1/\sqrt{2} \sqrt{12,800} = \sqrt{6400} = 80$ ft/sec. We now obtain the height above the starting point as a function of time:

$$y = v_y t - \tfrac{1}{2} g t^2.$$

We know that $t = \dfrac{360}{v_x} = \dfrac{360}{80} = \dfrac{9}{2}$ sec, when the ball is 360 ft from home plate

$$\therefore y = 80 \times \tfrac{9}{2} - \tfrac{1}{2} \times 32 \times \left(\tfrac{9}{2}\right)^2 = 360 - 324 = 36 \text{ ft}$$

∴ the ball is $4 + 36 = 40$ ft above the ground and the fence must have a height less than 40 ft.

3-5. Circular Motion. An object moving in a circular path with constant speed may not seem, at first thought, to be accelerated, but one must remember that velocity, as a vector quantity, is specified by both a magnitude and a direction. In the case of circular motion at constant speed, it is the direction which is continually changing. In this section we will derive an expression for the magnitude and direction of the acceleration associated with circular motion, using the fact that acceleration is defined as the rate of change of velocity with time, and that in general both changes in magnitude and direction must be considered. If, as in this case, the speed is constant, we need only worry about the effect of the changing direction.

In Figure 3-9(a) an object moving with constant speed in a circular path is shown at two positions on the circle. We ask what

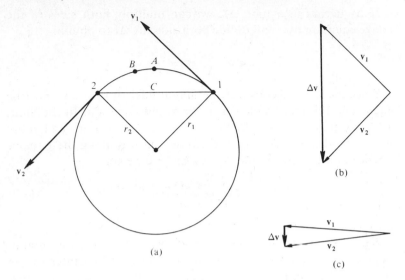

Figure 3-9. (a) The instantaneous velocity of an object moving with constant speed in a circular path at two positions. (b) The relationship between the instantaneous velocity at these two positions and the change in velocity, **Δv**. (c) Similar to (b), but for two points such as *A* and *B* which are much closer together as shown in (a).

vector must be added to v_1 to give us v_2. This must be the change in velocity experienced by the object as it goes between the two positions and this change, and its relationship to v_1 and v_2 is shown in Figure 3-9(b).

If one considers two points on the circular path which are very close together, then it should be clear, as indicated in Figure 3-9(c), that **Δv** is nearly perpendicular to v_1. Clearly, as Δt approaches zero, the angle between **Δv** and v_1 approaches 90° and the instantaneous $\Delta v/\Delta t$ must be directed toward the center of the circle.

The acceleration is defined as the limit of $\Delta v/\Delta t$, and to calculate this quantity we utilize Figure 3-9(a) and (b). Note that the triangle formed by r_1, r_2 and the cord C is similar to the triangle formed by v_1, v_2 and **Δv**, since v_1 is perpendicular to r_1, and v_2 is perpendicular to r_2. Therefore,

$$\frac{\Delta v}{v} = \frac{C}{r}.$$

If Δv occurs in a time Δt, we can multiply both sides of the above equation by v and divide both sides by Δt to obtain:

$$\mathbf{a} = \frac{\Delta v}{\Delta t} = \frac{Cv}{r\Delta t}.$$

Since the speed is constant the magnitudes of $\mathbf{v}_1$ and $\mathbf{v}_2$ must be equal, and this magnitude is $\mathbf{v}$ in the above equation. In the limit, as Δt is made very small, the cord C becomes more and more nearly the distance Δs which the object moves along the circumference of the circle in time Δt. As $\Delta t \to 0$ we get:

$$a = \frac{\Delta v}{\Delta t} = \frac{(\Delta s)v}{\Delta tr} = \frac{v^2}{r} \tag{3.2}$$

since $\Delta s/\Delta t = v$.

The quantity v^2/r is known as the centripetal acceleration and, as pointed out above, must be directed toward the center of the circle. Any object moving in a circular path with constant speed must experience this acceleration. If the speed is not constant, the acceleration may be broken up into radial and tangential components. The radial component will always be v^2/r, where v is the instantaneous speed. The tangential component will simply be the rate of change of speed, and, of course, the resultant acceleration will no longer be directed toward the center of the circle.

Example 3-9. With what speed must a toy train move in a circular path of radius 1 m if its centripetal acceleration is to equal $\frac{1}{20}$ the acceleration due to gravity?

Solution. Since $a = \dfrac{g}{20} = \dfrac{9.8}{20} = v^2/r$ we get,

$$v = \sqrt{ar} = \sqrt{0.49 \times 1} = 0.7 \text{ m/sec}.$$

Example 3-10. A car moving at the constant speed of 30 ft/sec makes a 90° turn in 3 seconds. (a) What is the average acceleration experienced by the car, calculated using only the above information? (b) Suppose that the car made the 90° turn while moving at the same constant speed in a circular path. What must the radius of this path be? What is the instantaneous acceleration experienced by the car in this path? Since the magnitude of this instantaneous acceleration is constant, why is it not equal to the result of part (a)? What would the result of part (a) be if the angle were 360° instead of 90°?

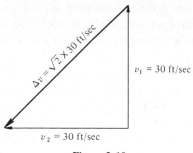

Figure 3-10.

Solution. (a) As indicated in Figure 3-10, the change in velocity must be $\Delta v = \sqrt{2} \times 30$ ft/sec, since Δv forms the hypotenuse of a right triangle the other two sides of which are vectors of length 30 ft/sec. Since $a = \Delta v/\Delta t$ must be the average acceleration, we obtain

$$a = \frac{30 \times 1.414}{3} = 14.14 \text{ ft/sec}^2.$$

(b) In turning through 90° the car must travel a distance of 30 ft/sec $\times$ 3 sec = 90 ft. Since it moves along the quadrant of a circle, 90 ft = $(\pi/2)\,r$ and $r = 90$ ft $\times$ 2/3.14 = 57.3 ft. The instantaneous acceleration for this path is given by $v^2/r = 900/57.3 = 15.7$ ft/sec^2. The result is not equal to the average acceleration calculated in part (a) because the average and instantaneous acceleration in a case such as this are quite different quantities. For example, if the angle in part (a) had been specified as 360° instead of 90°, the average acceleration as calculated in (a) would have been zero. Only if Δt, and therefore Δv, are reasonably small, will the average value approach the instantaneous value—as should be clear from the derivation of equation 3.2. If we were to repeat the calculation of part (a) for 45° and a time of 1.5 seconds (half the original value), and for 22.5° and 0.75 sec (a quarter the original value), we would obtain values of the average acceleration of 15.3 ft/sec^2 and 15.6 ft/sec^2, indicating the fairly rapid approach toward the instantaneous value as Δv gets smaller and smaller.

Example 3-11. A roller coaster car does a vertical loop-the-loop on a track which has a radius of curvature of 25 ft. What speed must the car have at the very top of the loop if its centripetal acceleration is to equal the acceleration due to gravity, g?

Solution. Using equation 3.2, we get $v = \sqrt{ar} = \sqrt{32 \times 25} = 28.28$ ft/sec.

Example 3-12. A racing car enters a circular turn on a race track. The radius of curvature of the turn is 200 ft, and at one particular point, as indicated in Figure 3-11, the instantaneous velocity of the car is 60 ft/sec, and the acceleration is -9 ft/sec^2, i.e., the car is slowing down. Determine the magnitude and direction of the resultant acceleration experienced by the car.

Solution. As indicated in Figure 3-11, the resultant acceleration is the vector sum of the centripetal acceleration and the tan-

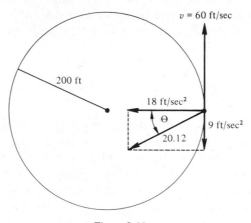

Figure 3-11.

gential acceleration. The centripetal acceleration is given by $a = v^2/r = (60)^2/200 = 18$ ft/sec^2, while the tangential acceleration is given as 9 ft/sec^2. The resultant acceleration is

$$a_R = \sqrt{18^2 + 9^2} = 20.1 \text{ ft/sec}^2,$$

making an angle $\theta = 26.6°$ with the radial direction. (Tan $\theta = 1/2$.)

3-6. Angular Velocity and the Radian. Usually angles are measured in terms of the degree, with 360° in a complete circle and 90° in the familiar right angle. There are many advantages, however, in defining a unit of angular measure in terms of the radius of a circle. This unit is called the radian and we define it with

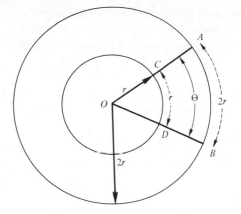

Figure 3-12. The angle Θ is one radian, since the arcs *CD* and *AB* are equal in length to one radius.

the help of Figure 3-12. The radian is defined so that one complete circle contains 2π radians. Since the circumference of a circle is given by $2\pi r$, one radian must subtend an arc length of one radius. The angle Θ in Figure 3-12 is one radian, and cuts out arc lengths *CD* and *AB* on circles of radius *r* and *2r*, respectively. Thus, 360° must equal 2π radians and 1 radian = $360/2\pi$ = $360/6.28$ = $57.296°$.

The rate at which an object rotates is expressed in terms of its angular velocity. This velocity can be expressed in terms of revolutions per minute, degrees per second, or radians per second. The latter unit, using the radian is particularly convenient, because it greatly facilitates the determination of the speed with which any part of a rotating object is moving. For example, suppose that Figure 3-12 represents a wheel rotating about its center at *O*. If, at a particular instant, the angular velocity is 3 radians per second, this must mean that in one second the point *C* moves a distance *3r* along a circle of radius *r*, and that the point *A* moves a distance *6r* along a circle of radius *2r*. Thus, the relationship between the angular velocity ω, in radians per second, and the linear speed of a point at a distance *r* from the axis must be

$$v = \omega r. \tag{3.3}$$

Example 3-13. A stone on the end of a string moves in a horizontal circle of radius 2 meters at an angular speed of 10 radians per second. What is the linear speed, in meters per second, with which the stone moves?

Solution.   We use equation 3.3 directly, obtaining:

$$v = \omega r = 10 \times 2 = 20 \text{ m/sec.}$$

Example 3-14.   Obtain an expression for the centripetal acceleration for an object moving in a circular path of radius r with angular velocity ω.

Solution.   Since the centripetal acceleration is given by $a = v^2/r$ and $v = \omega r$, we get $a = \omega^2 r^2/r = \omega^2 r$.

Chapter 4

Dynamics

4-1. Introduction. Using the concepts of displacement, velocity, and acceleration, we have seen that it is possible to describe many types of motion without worrying about how these motions are produced. In this chapter we will introduce the concept of force and investigate the relationship between the forces which act on a body and the motion which results from the action of these forces. This area of mechanics is known as dynamics.

Although we have used the term force without defining it, the everyday concept of a force as a push or pull in a particular direction is quite close to the definition which we shall develop for force. Obviously, the effect of a force on an object will depend on its direction. For example, if one pushes straight down on a suitcase resting on the floor, it will not move, but if the force is directed parallel to the floor and is large enough, the suitcase will slide across the floor. Clearly, it is necessary to specify both the magnitude and the direction of a force and thus we should not be surprised to find that it is a vector quantity. Therefore, when more than one force acts on a body, the resultant force must be determined by combining the forces according to the rules of vector addition. A more precise definition of force and the units of force is based on Newton's second law and will be given in the following sections.

4-2. Newton's Laws of Motion. Newton's laws of motion, which we discuss in this section, appeared as part of his *Philosophiae Naturalis Principia Mathematica*, published in 1686. This was one of the most important scientific documents in the history of science. In it, Newton collected and properly interpreted the work of generations of scientists before him, and in so doing essentially established the science of mechanics as we know it today.

Newton's three laws are presented in much the same form as used by Newton three hundred years ago, even though the first law is clearly a special case of the second. Despite this redundancy, the laws are in a useful form for application to the many kinds of problems which arise in the field of mechanics.

Newton's first law states that a body at rest, or moving in a straight line with constant velocity, will remain in that condition unless acted upon by an unbalanced force. Therefore, one concludes that if a body is at rest or moving in a straight line with constant velocity, the vector sum of all forces acting on it must be zero. Although these ideas seem simple and reasonable enough to us today, the problem of motion and its causes was studied by philosophers for centuries and until the time of the great Italian scientist, Galileo Galilei (1564–1642), most people believed that some sort of "force" was necessary to keep a body in motion since the "natural state" of a body was supposedly at rest.

Newton's second law states that the acceleration experienced by an object is directly proportional to, and in the same direction as the unbalanced force (i.e., the vector sum of all the forces) on the object, and inversely proportional to the mass of the object. This basic relationship between force, mass, and acceleration can justifiably be called the foundation of mechanics.

A simple example might be useful to emphasize the significance of Newton's second law. Let us consider three sleds on a smooth frozen pond for which, to a good approximation, there is no friction. The sleds are loaded with bricks with a different number of bricks on each sled. If each sled is pushed across the ice with the same constant unbalanced force and we observe an acceleration a_1 for the first sled, $a_1/2$ for the second sled, and $a_1/3$ for the third sled, we can say that the total mass of the second sled is two times that of the first, while the mass of the third sled is three times that of the first. This, of course, is because the acceleration is inversely proportional to the mass. We have therefore devised a method by which, in principle, we might compare one mass with another. If we were to pick one of the sleds and exert forces which gave accelerations of a, $2a$ and $3a$, we could conclude that the forces used had magnitudes of 1, 2, and 3 relative to the first, and that we have found a method of comparing forces in terms of the accelerations that they produced.

Newton's third law states that when two objects interact in such a way that one of them exerts a force on the second, the second exerts an equal and opposite force on the first. From this we conclude that forces always occur in pairs. If one force is called the action, and the other the reaction, it should be clear that action and reaction can *never* act on the same body. For example, a man exerting a force **F** on a wall experiences an equal and opposite force exerted on him by the wall. We shall be concerned with applications of the first and third law in Chapter 5, which deals with forces acting on objects at rest.

4-3. Newton's Second Law and Systems of Units. At this point we require a precise, quantitative definition of the unit of force and Newton's second law provides the basis for this definition in terms of the standard kilogram, the standard meter, and the second. The second law, as stated above, may be written in mathematical form as $F = km\mathbf{a}$, indicating that the force is proportional to the product of the mass and the acceleration, k being a constant of proportionality. The absolute M.K.S. system (meter, kilogram, second) defines a unit of force, which is called the newton, in such a way that k in the above equation is one. This is easily done by defining the newton as that net force which produces an acceleration of 1 m/sec^2 when acting on a mass of one kilogram. The term absolute is used to describe this system to distinguish it from a gravitational system such as the English gravitational or engineering system described later. In the latter system the unit of force is defined in terms of the gravitational *force* on a standard object, such as the standard kilogram. In this system, the unit of mass, rather than the unit of force, is a derived unit, and is defined in terms of Newton's second law.

Our expression for the second law now becomes:

$$F = m\mathbf{a} \qquad (4.1)$$

in which F is measured in newtons, m in kilograms, and $\mathbf{a}$ in meters/sec^2. It should be emphasized that F is the vector sum of all the forces acting on the mass m, that F and $\mathbf{a}$ are vector quantities, and that the direction of $\mathbf{a}$ is the same as that of F.

Example 4-1. A boy exerts a constant force of 40 newtons on a box which he pushes along the floor. The box has a mass of 20 kilograms and experiences a frictional force opposing the motion, of 10 newtons. Calculate the acceleration of the box.

Solution. We use equation 4.1 directly, but must be sure to determine the vector sum of all the forces acting on the box. We need not concern ourselves with the force of gravity, which we discuss in the next section, because this force is balanced by the component of the force exerted on the box by the floor perpendicular to the floor. Here the unbalanced force is 40 − 10 = 30 newtons and,

$$a = \frac{F}{m} = \frac{30}{40} = 0.75 \text{ m/sec}^2.$$

Note that the newton has the units kg m/sec^2.

Although the M.K.S. system, for which the fundamental definitions are provided above, and a similar C.G.S. system based on the centimeter ($\frac{1}{100}$ meter), the gram ($\frac{1}{1000}$ kg), and the second, are widely used in scientific circles, most people in the United States and Great Britian deal in their everyday life with quite different units, in which the unit of force is the familiar pound. Therefore, it seems reasonable to develop the English gravitational system which is still widely used in some fields of engineering.

In this system the force of gravity on the standard pound is the unit of force. The standard pound is a block of metal kept at the office of the Exchequer in London. In the United States, the pound is defined as 0.4535925 times the *weight* of the standard kilogram. The unit of mass in this system is called the slug, and is defined in terms of the second law such that a mass of one slug acted on by a force of one pound experiences an acceleration of 1 foot per second per second. Using this definition we see that since an object having a mass of one slug falls freely under the force of gravity and experiences an acceleration of 32 ft/sec^2, the net force on it must be 32 lbs. This force, of course, is its weight, and thus we must divide the weight of an object in pounds by 32, and acceleration due to gravity, to obtain its mass in slugs, i.e., one slug has a weight of 32 lbs.

4-4. Weight and Mass. It is important that the distinction between weight and mass be very clear because these concepts will be used repeatedly in our discussion of mechanics. Weight, of course, is a force—the force of gravity on an object—and thus must be expressed in newtons or pounds. Newton's second law indicates that force and mass are directly proportional and this can lead to some confusion. Using equation 4.1 we see that,

$$\mathbf{w} = m\mathbf{g} \tag{4.2}$$

where **w** is the weight of an object of mass m and **g** is the acceleration due to gravity. This is the result of the experimental fact that all objects at the same location fall with the same acceleration under the force of gravity. Confusion arises because we often

compare masses by determining their weights. We do this when
we compare an unknown mass with a standard mass using a
balance. We also describe objects by sometimes giving their mass
and sometimes their weight. Usually this depends on the system
of units we are using. In the M.K.S. system it is common to give
the mass in kilograms but in the English system for example, we
almost always buy meat by specifying its weight in pounds and
not its mass in slugs. Using equation 4.2, and the fact that
$g = 32$ ft/sec^2 $= 9.8$ m/sec^2, we see that the mass of an object
weighing 1 lb. is 1/32nd slug, while the weight of a 1 kilogram
object is 9.8 newtons. Since our definition of the pound involved
the fact that a 1 lb object has the same weight as an object with
a mass of 0.4536 kg, we note that 1 kg has a weight of 2.2046 lbs.
Similarly, 1/32 slug = 0.4536 kg, so that 1 slug = 14.5149 kg, and
since 9.8 newtons is the weight of 1 kg, we find that 1 newton
$= 0.224$ lb.

From equation 4.2 it should be apparent that the weight of an
object will change from one location to another if g the acceleration due to gravity, changes. The mass, however, is the same at
all locations. The acceleration due to gravity changes from place
to place on the surface of the earth, but over the surface of the
earth varies by less than one percent, being about 9.832 m/sec^2
at the poles, and 9.780 at the equator. A considerably more dras-
tic change in g would result if one suddenly found himself on the
planet Mars, where the acceleration due to gravity is only one
third that on earth. If a 9 lb object were weighed on a spring
balance properly calibrated on earth, the spring would stretch a
certain amount and the dial would read 9 lbs. If the measurement
were repeated on Mars with the same 9 lb object, and the same
spring balance, the spring would stretch only one third as much,
and the balance would read 3 lb. If, however, the measurement
were made with an equal arm balance, which compares the force
of gravity on an unknown with the force of gravity on a standard
mass, the value obtained on Mars would be the same as that
measured on earth, because the force of gravity is changed by
the same factor for both standard and unknown.

Although the weight of an object such as a baseball would be
different on Mars, one would find that throwing it would be much
like throwing it on earth as far as the horizontal motion is concerned. One could throw further and higher—as suggested by the
material in Table 3-2—since g is less, but the force required to

obtain a given horizontal acceleration would be the same, and a pitcher on Mars would not be able to throw any faster than on earth. A batter, on the other hand, would be considerably better off, since a 300 ft fly ball on earth would result in a 900 ft home run on Mars. One might thus conclude that the prospects for baseball, as we know it, on Mars are rather slim.

4-5. Newton's Second Law and Problem Solving. We now proceed to illustrate the fundamental nature of Newton's second law by applying it to several numerical problems. Before doing this it is important to emphasize the general manner in which such problems should be approached. Obviously, in dealing with Newton's second law, one or more forces act on some object and that object experiences an acceleration. One should be careful to specify or "isolate" the object to be considered, and then to list all the forces acting on it and calculate the vector sum of these forces. With the exception of the force due to gravity, and electric and magnetic forces with which we are not concerned here, it is not possible to exert a force on an object without touching it. Therefore, it becomes simply a matter of listing the force due to gravity, deciding whether or not it is directly concerned with the problem at hand, and then observing all objects in contact with the body, listing the magnitude and the direction of the forces which they exert.

Example 4-2. A man moves upward in an elevator which has a constant acceleration of 6 ft/sec^2. With what force do his feet press against the floor of the elevator if he weighs 160 lbs?

Solution. Here it is reasonable to "isolate" the man and list all the forces acting on him. There are only two, these being the force of gravity acting downward, and the force exerted on the man by the elevator at his feet, which is directed upward. We designate this last force by **F** and realize that it is oppositely directed to his weight and larger than his weight. This must be true since the man is accelerating upward. Applying equation 4.1 we obtain:

$$\mathbf{F} - m\mathbf{g} = m\mathbf{a},$$

and since $m\mathbf{g} = 160$ lbs and $m = \dfrac{w}{g} = \dfrac{160}{32} = 5$ slugs

we may write:

$$\mathbf{F} = m\mathbf{g} + m\mathbf{a} = 160 + 6 \times 5 = 160 + 130 = 190 \text{ lbs}$$

which is the force exerted by the floor on his feet. This, of course, is equal and opposite, by Newton's third law, to the force exerted by his feet on the floor of the elevator.

Example 4-3. Consider a block sliding down an inclined plane which makes an angle of 37° with the horizontal, and is so smooth that we do not have to worry about friction (see section 4-6). Calculate the acceleration experienced by the block as it slides down the plane.

Solution. Our procedure is to "isolate the block and list all the forces acting on it." Here again there are only two: the force of gravity, $m\mathbf{g}$, and the force labeled **N** in Figure 4-1, which is exerted on the block by the plane and which must be perpendicular to the plane because of our assumption that the plane is smooth and there is no friction. The force of gravity can be broken up into components parallel and perpendicular to the plane, these being $mg \sin \Theta$ and $mg \cos \Theta$ as indicated in Figure 4-1. Since there is no acceleration in the direction perpendicular to the plane, the forces acting on the block in that direction must sum to zero. Therefore the force **N** must be equal and opposite to $mg \cos \Theta$, the component of the force of gravity perpendicular to the plane. Thus we can consider the component of the force of gravity parallel to the plane, $mg \sin \Theta$, as the only force acting in the direction of motion of the block, and apply Newton's second law to get: $a = F/m = mg \sin \Theta/m = g \sin \Theta$, indicating that the acceleration is independent of m and that one can obtain any

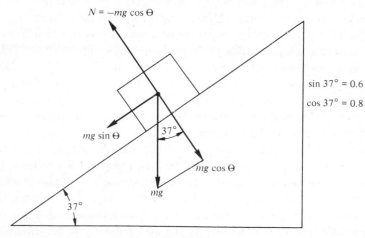

Figure 4-1.

desired acceleration between 0 and g for such a frictionless plane by appropriately choosing Θ.

Example 4-4. The device shown in Figure 4-2 is known as an Atwood's machine. The masses are connected by a cord of negligible weight and the pulley has negligible friction. Calculate the

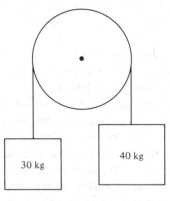

Figure 4-2. Atwood's Machine.

acceleration experienced by the system when free to move if one of the suspended objects has a mass of 40 kg and the other a mass of 30 kg. Also calculate the tension in the cable. The tension in the cable is defined as the force exerted by the cable on the object to which it is connected. Since the cable is massless, the effect of the cable is to transmit the force from one end to the other undiminished. In other words, the force exerted by the cable on both blocks is the same, and is equal to the tension in the cable.

Solution. Perhaps the most direct approach is to consider the system as a whole with a total mass of $30 + 40 = 70$ kg. The 40 kg mass moves down and the 30 kg mass moves up—both moving with the same acceleration. In this system, the force of gravity on one block opposes the force of gravity on the other. Therefore, the resultant or unbalanced force on the system is the difference between the two, which is $(40 - 30)g = 10 \times 9.8 = 98$ newtons. The acceleration of both masses should be given by $a = F/m = 98/70 = 1.4$ m/sec^2. In this approach the tension in the cable becomes an internal force and cancels out. The tension in the cord is obtained by isolating either one of the masses, as illustrated in Figure 4-3, listing all the forces on it, and applying Newton's second law. Choosing the 30 kg mass, we see that the only forces acting are gravity and the force T exerted by the cord.

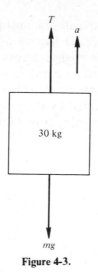

Figure 4-3.

Clearly T is larger than mg since the 30 kg mass moves upward. We calculate the vector sum of these forces, which is $T - mg$, directed upward, and set it equal to ma, which in this case is 30 kg × 1.4 m/sec². Thus,

$$T - mg = ma$$
$$T - 30 \times 9.8 = 30 \times 1.4$$
$$T = 30\,(9.8 + 1.4) = 336 \text{ newtons}$$

which is greater than the weight of the 30 kg object but less than the weight of the 40 kg object.

The procedure followed in calculating the tension above could have been used on both masses, obtaining two simultaneous equations, both involving the acceleration and the tension. Applying Newton's second law to both masses, these equations would be:

$$T - 30 \times 9.8 = 30a \quad \text{for 30 kg mass}$$

and $\quad 40 \times 9.8 - T = 40a \quad$ for 40 kg mass.

Solving for T in each equation gives:

$$T = 30\,a + 294$$

and $\quad T = 392 - 40\,a.$

Eliminating T gives $70\,a = 98$ and $a = 98/70 = 1.4$ m/sec², as obtained above. Note also that each equation, when solved for T, gives the value 336 newtons.

4-6. Friction. In almost all practical problems encountered in mechanics one cannot neglect friction as we did in Example 4-2. When a block of wood is pushed across a horizontal surface, for example, it is observed that a constant force is required to maintain its motion at constant speed. This force is required to overcome the resistance provided by irregularities in the two surfaces as illustrated (greatly magnified) in Figure 4-4(a). The force is

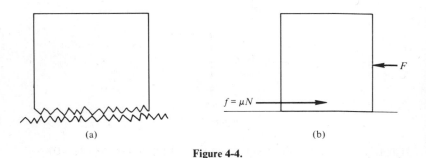

(a)   (b)

Figure 4-4.

required to push the block over these small hills and valleys or perhaps grind off the tops of the hills. With some surfaces it is also very likely that the intimate contact between the two objects results in an actual welding together of the two surfaces at the high spots. To maintain this constant making and breaking of many small joints requires a force which also contributes to the force of friction.

Although the exact nature of the force of friction is imperfectly understood, it almost certainly is due in a large part to some combination of the effects mentioned above. Experimentally it is found that for many situations some very simple laws can be used to describe frictional forces. It is found that *the frictional force is proportional to the normal force between the surfaces*—i.e., the force pressing the two surfaces together—and that *it is independent of the area of contact*, and, of course, depends on the nature of the surfaces. Mathematically these statements can be expressed as:

$$\mathbf{f} = \mu \mathbf{N} \qquad (4.3)$$

where $\mathbf{f}$ is the frictional force, $\mathbf{N}$ is the normal force and μ is a constant called the coefficient of friction which is determined by the nature of the surfaces, i.e., by how rough they are and what they are made of. In general, coefficients of friction tend to be less than one, but can, of course, be greater. Some pairs of surfaces, such as the plastic teflon on certain metals, can have a coefficient of friction as low as 0.05.

Electrical measurements have indicated that the fraction of any two surfaces actually in contact is very small, being of the order of 10^{-5}. This fact gives us some notion as to why the frictional force experimentally turns out to be very nearly independent of the apparent area of contact.

Let us consider a block at rest on a horizontal surface as indicated in Figure 4-4(b). If a horizontal force **F** is exerted on the block, it will not move until the force is equal to μN. Thus, the frictional force can be anything from zero up to a maximum of μN depending on the circumstances. It turns out that the coefficient of friction is slightly less for a pair of surfaces after motion has started, and it is customary to refer to coefficients of static and kinetic friction to account for this effect.

Example 4-5. A 50 kg block rests on a horizontal surface. (The coefficient of static friction for the two surfaces is 0.5.) A horizontal force of 200 newtons is exerted on the block. What is the magnitude and direction of the frictional force experienced by the block?

Solution. The normal force between the two surfaces is equal to the weight of the block, which is $50 \times 9.8 = 490$ newtons. The maximum frictional force is given by $f = \mu N - 0.5 \times 490 = 245$ newtons. Since the horizontal force is less than μN, the frictional force will be equal and opposite to the applied force of 200 newtons.

Example 4-6. Obtain an expression for the acceleration of a block sliding down an inclined plane of angle θ if the coefficient of friction between the plane and the block is 0.2. Show that the coefficient of static friction is given by $\mu = \tan \theta$, where θ is the maximum angle at which the block will not move. Show also that the coefficient of kinetic friction is given by the same expression, where θ is the angle at which the block slides at constant speed down the plane.

Solution. The two forces acting on the block are mg, the force of gravity and **F**, the force exerted on the block by the plane as indicated in Figure 4-5. If each is broken into components parallel and perpendicular to the plane we note that the normal force, N, is given by $mg \cos \theta$, and that the vector sum of the components

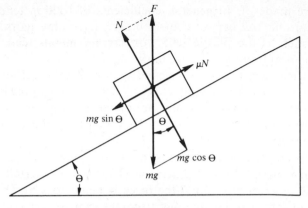

Figure 4-5.

parallel to the plane is $mg \sin \theta - \mu mg \cos \theta$. Thus, when we apply Newton's second law we get:

$$mg \sin \theta - \mu mg \cos \theta = ma$$

so that,

$$a = g(\sin \theta - \mu \cos \theta).$$

If a is zero, as it must be if the block is just on the verge of moving, or is moving with constant speed, we get

$$g \sin \theta = \mu g \cos \theta$$

and

$$\mu = \frac{\sin \theta}{\cos \theta} = \tan \theta.$$

The angle θ should be slightly greater for the case in which the object is still at rest, and for which the above equation yields the coefficient of static friction.

Example 4-7. Calculate the acceleration experienced by the two weights shown in Figure 4-6, if the coefficient of friction between the 32 lb weight and the plane is 0.2. Calculate also the tension in the cable whose weight we assume to be negligible.

Solution. As in Example 4-4, it is easiest to consider the motion of the system as a whole. We need only consider the force of gravity on both weights and the frictional force on the 32 lb weight. The frictional force is given by $f = \mu N = 0.2 \times 32 \cos \theta$, and is directed down the plane, since it opposes the motion. All

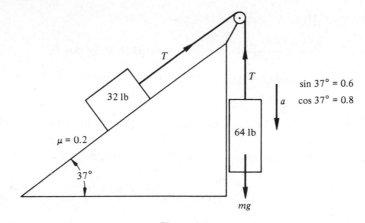

Figure 4-6.

the cable does is change the direction of the force which it transmits. Applying Newton's second law we get:

$$64 - 32 \sin \Theta - \mu \times 32 \cos \Theta = (M + m)a = \left(\frac{64}{32} + \frac{32}{32}\right)a$$

$$64 - 19.2 - 5.1 = 3a$$

$$39.7 = 3a$$

$$a = 13.2 \text{ ft/sec}^2.$$

The tension is obtained by isolating the 64 lb weight and noting that the only two forces acting on it are T and the force of gravity. Calling the downward direction positive we obtain, from Newton's second law:

$$64 - T = Ma = \frac{64}{32} \times a = 2a = 26.4$$

$$T = 64 - 26.4 = 37.6 \text{ lbs.}$$

Example 4-8. A block is given an initial velocity of 20 ft/sec up a 20° inclined plane. How far up the plane will the block slide if the coefficient of friction between block and plane is 0.3. The sin of 20° is 0.3420 and the cos of 20° is 0.9397.

Solution. We first obtain the acceleration, which is directed down the plane, and results from the component of the force of gravity parallel to the plane and the force of friction in the same direction. Newton's second law gives:

$$a = \frac{mg \sin \Theta + \mu mg \cos \Theta}{m} = g \sin \Theta + 0.3g \cos \Theta$$

$$a = 32 \times 0.342 + 0.3 \times 32 \times 0.9397 = 10.9 + 9.0$$

$$a = 19.9 \text{ ft/sec}^2.$$

The velocity, distance, and acceleration are related by equation 2.7 which gives $v^2 = v_0^2 - 2ad$, and when the block reaches its highest point and stops, $v = 0$, so that $d = v_0^2/2a$, and

$$d = \frac{(20)^2}{(2 \times 19.9)} = \frac{400}{39.8} = 10.1 \text{ ft.}$$

4-7. Circular Motion and Centripetal Force. In Chapter 3 we found that all objects which move with velocity v in a circular path of radius R must experience an acceleration, directed toward the center of the circle, of magnitude v^2/R. It should be clear from our discussion of Newton's second law that a force is required to produce this acceleration. This force is easily calculated using the second law in the form $F = ma$ and substituting for a the centripetal acceleration, v^2/R obtaining:

$$F = \frac{mv^2}{R}. \tag{4.4}$$

This force is called the centripetal force and must be directed toward the center of the circle. Any object moving in a circular path must experience an unbalanced force of this magnitude.

It is a common experience to ride in an automobile or train rounding a curve and feel an apparent force pushing one away from the center of the circular path. This force is commonly called centrifugal force, but a moment's thought indicates that no such force can exist. This terminology has developed because our senses tell us that we are being forced toward the outside of the curve, but what actually is happening, expressed in terms of Newton's laws, is that the passenger tends to move at constant speed in a straight line. In rounding the corner, the car or train deviates from this straight line, carries the passenger with it and, in so doing, tends to move out from under him, carrying him along because of the friction between the car seat and the seat of the passenger's pants. This frictional force provides the centripetal force to move the passenger along the curved path and the force on the passenger is toward the center of the circle and not radially outward. The reaction to this centripetal force—in our example the force exerted by the passenger on the car—is directed outward, and it is this force that some scientists call the centrifugal force. Note again that it is not acting on the passenger. Despite our emphasis that no "centrifugal force" is exerted on the passenger, the effect is a very real one to him. In many problems in mechanics, it is convenient to assume the existence of such a force, while realizing, of course, that it is fictional. Imagine, for example, the situation that would exist if one were to live for a while on a large, rotating merry-go-round. In addition to the force of gravity, one would also experience an *apparent* horizontal force directed radially outward. Relative to the merry-go-round, a ball on the floor would always roll outward, and falling objects would not drop straight down but would fall at an angle. Hence, in such a rotating coordinate system, the apparent "centrifugal force" on all objects is a very real one.

Example 4-9. A 64 lb block moves in a horizontal circle of radius $R = 3$ ft on a smooth surface for which friction is negligible, as shown in Figure 4-7. Calculate the tension in the cord

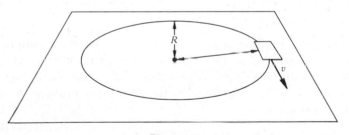

Figure 4-7.

connecting the rotating block to the center of the circle if the block moves with a speed of 10 ft/sec.

Solution. The cord provides the centripetal force which must be given by $\dfrac{mv^2}{R}$. Therefore $m = \dfrac{w}{g} = \dfrac{64}{32} = 2$ slugs ,

$$\frac{mv^2}{R} = 2 \times \frac{(10)^2}{3} = 66.7 \text{ lbs.}$$

Example 4-10. Figure 4-8 illustrates the motion of a conical pendulum—a mass on the end of a cord of length L—which

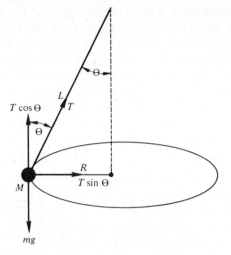

Figure 4-8.   A conical pendulum.

moves in a horizontal circle of radius R, the cord tracing out a conical surface in space as a result of the motion. Obtain an expression for the angle Θ in terms of the length of the pendulum, L, and the angular frequency $\omega = 2\pi f$. Show also that the tension in the cord is given by $T = m\omega^2 L$.

Solution.   As indicated in Figure 4-8, the vertical component of the force exerted on the mass by the cord must be equal and opposite to the force of gravity, and the horizontal component of the force exerted on the mass by the cord must provide the centripetal force so that:

$$T \sin \Theta = \frac{mv^2}{R} = m\omega^2 R,$$

since $v = \omega R$ and

$$T \cos \Theta = mg.$$

Since $\sin \Theta = R/L$ we get:

$$T \sin \Theta = m\omega^2 L \sin \Theta$$

and

$$T = m\omega^2 L.$$

Substituting this value of T in the equation $T \cos \Theta = mg$, we obtain:

$$\cos \Theta = \frac{mg}{T} = \frac{mg}{(m\omega^2 L)} = \frac{g}{(\omega^2 L)}.$$

Curves on roads and railroads are often banked to facilitate high speed turning. Figure 4-9 illustrates the effect of the bank-

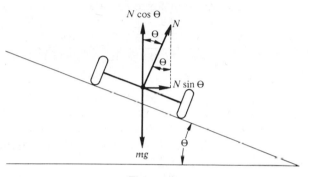

Figure 4-9.

ing of a curve. The forces acting on the car are its weight, $m\mathbf{g}$, and the force $\mathbf{N}$ exerted on the car by the road. If the curve is properly banked there is no frictional force on the tires in a sideways direction, and $\mathbf{N}$ is perpendicular to the road. The required centripetal acceleration of the car is provided by the horizontal component, $N \sin \Theta$. The vertical component, $N \cos \Theta$, just balances the downward force, mg. From this we see that N is greater than mg, since $N = mg/\cos \Theta$. A passenger in the car will feel that the seat is pushing up on him a little harder than usual, but at an angle Θ to the vertical, so that he experiences no tendency to move toward the outside of the curve. For a properly banked curve we have found that

$$N \sin \Theta = \frac{mv^2}{R} \quad \text{and} \quad N \cos \Theta = mg.$$

Dividing the first equation by the second we obtain:

$$\frac{\sin \Theta}{\cos \Theta} = \tan \Theta = -\frac{v^2}{Rg}. \tag{4.5}$$

Equation 4.5 shows very clearly that for one speed and one radius of curvature there is only one angle which satisfies the requirements for proper banking of a curve.

Example 4-11. A racing car moving at a speed of 100 mph (146.7 ft/sec) moves around a curve on a race track of radius of curvature 300 ft (the length of a football field). At what angle must the curve be banked in order that there be no tendency for the wheels to slip sideways?

Solution. We use equation 4.5 to calculate tan Θ.

$$\tan \Theta = \frac{v^2}{Rg} = \frac{(146.7)^2}{(300 \times 32)} = \frac{21511}{(9600)}$$

$$\tan \Theta = 2.24 \quad \text{and} \quad \Theta = 66°.$$

Example 4-12. A car in a toy road racing set must go through a loop the loop of radius 1 ft. How fast must this car be going at the very top of the loop, as indicated in Figure 4-10, if the

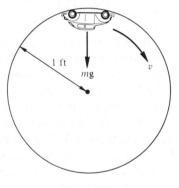

Figure 4-10.

car is to "just make it" around the loop with the wheels touching the loop but exerting negligible force on it?

Solution. The minimum velocity possible at the top of the loop with the car still in contact with the track must be such that the force of gravity provides the required centripetal force, i.e., $mg = mv^2/R$. This means that:

$$\frac{v^2}{R} = g, \quad \text{and} \quad v = \sqrt{\frac{g}{R}} = \sqrt{\frac{32}{1}} = 5.66 \text{ ft/sec.}$$

Example 4-13.   An airplane, in pulling out of a vertical dive, moves along the arc of a circle of radius 500 ft (the plane of the circle being vertical). If the speed of the plane is 200 mph (293.3 ft/sec) at the bottom of this loop, what force is exerted on the 160 lb pilot by the plane, and how many times greater than the normal force of gravity is this, i.e., how many "g's" does the pilot experience?

Solution.   As indicated in Figure 4-11, at the bottom of the loop there are two forces, oppositely directed, which act on the

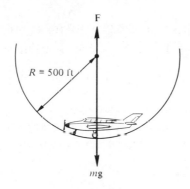

$R = 500$ ft

Figure 4-11.

pilot   the downward force of gravity, $m\mathbf{g}$, and the upward force $\mathbf{F}$ exerted on him by the plane. Since we know that at the bottom of the loop the pilot is moving in a circular path of radius 500 ft, the net or unbalanced force on him must be mv^2/R directed upward toward the center of the circle. Therefore:

$$F - mg = \frac{mv^2}{R} \quad \text{and} \quad F = mg + \frac{mv^2}{R} = m\left(\frac{g + v^2}{R}\right)$$

$$\frac{v^2}{R} = \frac{(293.3)^2}{500} = \frac{86042}{500} = 172.1$$

and

$$F = m(32 + 172.1) = 204.1m = \frac{204.1}{32} mg = 6.38mg.$$

There the force exerted by the plane on the pilot is 6.38 times his normal weight of 160 lbs or 1020 lbs.

Example 4-14.   Calculate the velocity which an earth satellite must have if it is to move in a circular orbit fairly close to the surface of the earth. Assume that the earth is a sphere of radius 3960 miles. Neglect friction due to the atmosphere but note that our answer would be only slightly different if we added 100 miles to the radius so as to move the satellite out beyond most of the atmosphere.

Solution. The centripetal force required to keep the satellite in its circular orbit must be provided by the force of gravity, *mg*. This gives us

$$mg = \frac{mv^2}{R} \quad \text{or} \quad v = \sqrt{Rg}$$

$$R = 3960 \text{ miles} \quad \text{or} \quad 20.9 \times 10^6 \text{ ft} \quad (1 \text{ mile} = 5{,}280 \text{ ft})$$

$$v = \sqrt{20.9 \times 10^6 \times 32} = 2.59 \times 10^4 \text{ ft/sec}$$

$$v = 2.59 \times 10^4 \times \frac{60}{88} = 1.76 \times 10^4 \text{ mph}$$

since 60 mph corresponds to 88 ft/sec.

In Chapter 8 we discuss universal gravitation and planetary motion and there we shall have more to say about satellites.

Chapter 5

Statics

5-1. Introduction. There are many problems that arise in the various areas of science and engineering that do not involve motion, but that are concerned with the effects of forces acting on bodies which are at rest. These problems involve the calculation of such things as the tension in a cable or the tension or compression in a particular part of a rigid structure. We shall first discuss the class of problems which deal only with situations in which all of the forces acting on an object pass through the same point. We then move on to the slightly more complicated situation for which this is not the case. In the latter situation the forces have a tendency to twist or rotate the body on which they act. Problems of these types constitute the field of statics and are solved using Newton's first and third laws.

5-2. Equilibrium—Concurrent Forces. Recall that Newton's first law states that an object must be at rest or in motion with constant velocity if the vector sum of all the forces acting on it is zero. One way of conveniently expressing this fact, in two dimensions, is to require that the sum of the x components and the sum of the y components of all the forces must be zero. This, of course, says nothing new, but it is a useful way of looking at things when dealing with many practical problems. We express this condition mathematically as:

$$\Sigma F_x = 0 \qquad \Sigma F_y = 0 \qquad (5.1)$$

where the Greek letter sigma (Σ) means "the sum of." Equation 5.1 is often called the first condition for equilibrium, and if this condition is satisfied (for concurrent forces), a body at rest will remain at rest with no tendency to move in a straight line or to rotate.

Let us apply equation 5.1 and Newton's third law to the rather simple situation illustrated in Figure 5-1. This figure shows a 100 lb weight suspended from the ceiling by a rope of negligible mass. The weight is at rest, and therefore it must be in equilibrium, and the vector sum of all the forces acting on it must be zero. The force exerted on the weight by the rope is directed upward,

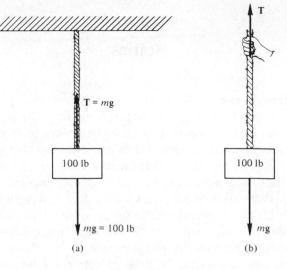

Figure 5-1.

and the force of gravity is down, so that T, the tension in the rope, must be equal in magnitude to mg and oppositely directed, which is not hard to believe. Note that if we detach the weight from the ceiling and have a man pull upward with a force of 100 lbs on the rope, we now have the weight pulling down on the rope with a force of 100 lbs, and the man pulling up with a force of 100 lbs. What is the tension in the rope under these circumstances? Is it 200 lbs? The answer is, of course, that nothing has changed; the man has merely replaced the ceiling which, by

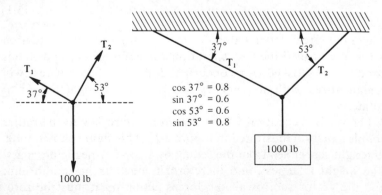

Figure 5-2.

Newton's third law, also exerted a force of 100 lbs on the rope, and the tension is still 100 lbs.

Example 5-1. Consider the 1000 lb weight suspended by two weightless cords as shown in Figure 5-2. Calculate the tension in all three cords.

Solution. It is convenient to isolate the "knot" where the three cords join. The forces acting at this point are shown in the vector diagram of Figure 5-2. It should be clear that the tension in the cord supporting the weight directly must be 1000 lbs. We apply equation 5-1, noting that the x and y components of the tensions T_1 and T_2 must be:

$$T_{1x} = T_1 \cos 37° = -0.8 T_1$$
$$T_{1y} = T_1 \sin 37° = 0.6 T_1$$
$$T_{2x} = T_2 \cos 53° = 0.6 T_2$$
$$T_{2y} = T_2 \sin 53° = 0.8 T_2.$$

Since $\Sigma F_x = 0$ we must have $0.8 T_1 = 0.6 T_2$ and $T_1 = 0.75 T_2$.

Since $\Sigma F_y = 0$ we must have $0.6 T_1 + 0.8 T_2 = 1000$ lbs.

Eliminating T_1 from the above equation we get:

$$0.6(0.75 T_2) + 0.8 T_2 = 1000$$
$$1.25 T_2 = 1000$$
$$T_2 = 800 \text{ lbs}$$

and

$$T_1 = 0.75 \times T_2 - 600 \text{ lbs}.$$

The student should note that the length of the cord plays no role. Only its inclination to the horizontal is of significance. If, for example, the left hand cord were shortened by attaching it to a nearby wall (not shown) the result would not be changed if the inclination of 37° were maintained.

Example 5-2. A 200 lb man hangs from the middle of a tightly stretched rope so that the angle between the rope and the horizontal direction is 5°, as shown in Figure 5-3. Calculate the tension in the rope.

Solution. The symmetry of the force diagram of Figure 5-3 indicates that the tension is the same in both sections of the rope. Thus we need only one of equations 5.1. $\Sigma F_x = 0$ is trivial, so we must use $\Sigma F_y = 0$. This gives us $200 = T \sin 5° + T \sin 5° = 2 \times T \times 0.0871$, and $T = 100/0.0871 = 1150$ lbs. Note the significant increase in force produced by this arrangement. The ten-

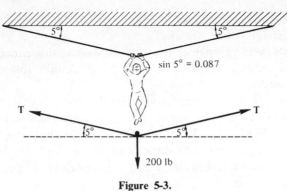

Figure 5-3.

sion in the rope is over 5 times the weight of the man. Had the angle been as small as 1° the tension would have been $T = 100/0.0174 = 5730$ lbs. This technique for exerting a large force would only be useful, of course, to move something a very small distance, since any motion of one end of the rope or cable would change the small angle considerably.

Example 5-3. A 1000 lb weight is suspended from the wooden boom (see Figure 5-4) whose weight we neglect. Calculate the tension in the supporting cable, and the compression in the boom (which may be taken to be along the boom since the weight of the boom is assumed to be small). This compression is the force exerted by the boom on the wall and the force exerted by the boom at the point of connection of the two cables.

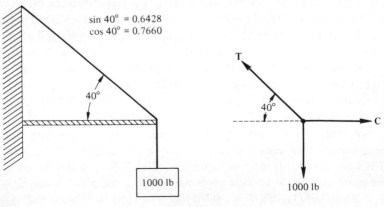

Figure 5-4.

Solution. We isolate the point at the end of the boom through which all three forces pass. The forces acting at this point are shown in the vector diagram of Figure 5-4. We apply equation 5.1 and get:

$$\Sigma F_y = 0 \rightarrow T \sin 40° = 1000 \text{ lbs, so that } T = \frac{1000}{0.6428}$$

$$T = 1556 \text{ lbs}$$

$$\Sigma F_x = 0 \rightarrow T \cos 40° = C,$$

where C is the compression in the boom. Therefore, $C = 1556 \times 0.7660 = 1192$ lbs.

5-3. The Second Condition for Equilibrium—Torques. When all the forces acting on a body do not go through the same point, it should be apparent that in general each of them tends to cause the object to rotate. Consider the plank shown in Figure 5-5. We neglect, for the present, the weight of the plank, and consider the forces exerted by the two supports and the 100 lb weight. We measure the tendency of a force to cause an object to rotate about a particular axis by multiplying the force by the perpendicular distance from the line of action of the force to the axis. This product is called the moment of the force about that axis, or, simply, the torque. In order that an object be at rest and remain at rest—i.e., to be in equilibrium—the sum of the torques as well as the sum of the forces must be zero. Torque is a vector quantity, the direction of the torque by convention being perpendicular to the plane in which the force acts, and is determined by a right hand rule such that if the fingers of the right hand curl in the direction of the rotation, the thumb, if extended, points in the

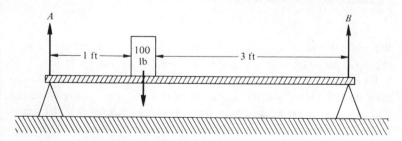

Figure 5-5. The plank experiences three forces but is in equilibrium because the sum of the torques as well as the sum of the forces on it are equal to zero.

direction of the torque, i.e., parallel to the axis about which the force tends to rotate the object on which it acts. For example, the torque of force B about an axis through the far end of the bar, and perpendicular to the paper, is directed out of the paper. Such a torque is conveniently described as being "counterclockwise" with its magnitude expressed in pound feet or newton meters. The 100 lb weight produces a clockwise torque of $100 \times 1 = 100$ pound feet about the left hand point of support. The force A produces zero torque about this point, since its line of action passes through the point. The fact that the sum of the torques must be zero on any body in equilibrium is often referred to as the second condition of equilibrium. The torques may be computed about any axis at all, on or off the object under consideration, but it is usually convenient to pick an axis on the object and also one through which an unknown force passes so as to eliminate this force from the resulting equations.

Example 5-4. Determine the forces A and B in Figure 5-5.

Solution. Since the plank is clearly in equilibrium we may utilize both conditions of equilibrium. There are no forces in the horizontal direction. $\Sigma F_y = 0$ yields:

$$A + B = 100 \text{ lbs.}$$

If we take moments about the left point of support through which the line of action of A passes we get:

$$\begin{array}{cc} \textit{Clockwise} & \textit{Counterclockwise} \\ 100 \times 1 = & B \times 4 \\ 100 = & 4B \end{array}$$

and

$$B = 25 \text{ lbs.}$$

Since $A + B = 100$,

$$A = 75 \text{ lbs.}$$

We can obtain A directly, of course, by taking moments about the right point of support, giving $4A = 300$, and $A = 75$ lbs.

Example 5-5. At what point should a uniform board 100 cm long be supported so that it balances if a 10 gram mass is placed on one end, a 60 gram mass on the other end, and a 40 gram mass 30 cm from the 10 gram mass (see Figure 5-6).

Solution. If the board is to balance, the sum of the moments about the point of support must equal zero. Let A be the distance from this point of support to the 60 gram mass. Equating clock-

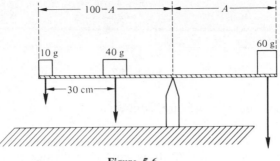

Figure 5-6.

wise and counterclockwise moments we get: (g is the acceleration due to gravity)

Clockwise	Counterclockwise

$$60 \times g \times A = 40 \times g \times (100 - A - 30) + 10 \times (100 - A)g$$
$$60A = 4000 - 40A - 1200 + 1000 - 10A$$
$$110A = 3800$$
$$A = 34.5 \text{ cm.}$$

5-4. Center of Gravity. The weight of the plank was neglected in example 5-4, but this cannot always be done. The force of gravity is actually distributed over the whole plank, but there is one point for any object through which the weight of the whole object may be considered to act. This point is called the center of gravity. The basis for calculating the location of the center of gravity is the fact that if the moments of the gravitational forces acting on all of the particles in the object are taken about the center of gravity, the sum would be zero. For many objects such as spheres, cylinders and rectangles of uniform composition, the center of gravity is at the geometrical center. Had we considered the weight of the plank in Example 5-4, it would only have been necessary to add another force equal to the weight of the plank and acting downward at the center of the plank. A moment's thought will show that the center of gravity does not have to be inside the solid matter composing an object. The center of gravity for many curved objects, such as a hemispherical shell, must lie outside the shell itself.

Example 5-6. Calculate the center of gravity of the object shown in Figure 5-7, assuming that the masses shown are essen-

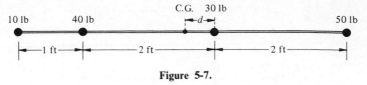

Figure 5-7.

tially point masses and are connected by a rod of negligible weight.

Solution. We assume that the center of gravity is some distance d to the left of the 30 lb weight. That it must be to the left, rather than the right, of the 30 lb weight can be seen by assuming that the 30 lb weight is located exactly at the center of gravity. If this were the case, the clockwise moments about the center of gravity would be 100 lb ft due to the 50 lb weight, while the counterclockwise moments due to the two weights on the left would be $40 \times 2 + 10 \times 3 = 110$ lb ft. Thus, the center of gravity must be some distance to the left of the 30 lb weight. Equating clockwise and counterclockwise moments about the center of gravity we get:

$$\textit{Clockwise}$$
$$50(2 + d) + 30d = 40(2 - d) + 10(3 - d)$$
$$80d + 100 = 110 - 50d$$
$$130d = 10$$
$$d = 0.077 \text{ ft} = 0.924 \text{ inches,}$$

and the center of gravity is 2.077 ft from the right end of the object.

Example 5-7. Solve Example 5-3 (page 60), assuming that the boom has a weight of 200 lbs. Find the force exerted on the boom by the wall.

Solution. Figure 5-8 shows all the forces acting on the boom and indicates that the force exerted on the boom by the wall does not act along the boom if the weight of the boom is not neglected. We have broken this force up into x and y components. Although it may not be immediately obvious that the y component is directed upward, it is seen that this must be the case by qualitatively taking moments about the right end of the boom. R_y must be directed upward to provide the necessary clockwise torque to maintain equilibrium. Using the second condition for equilib-

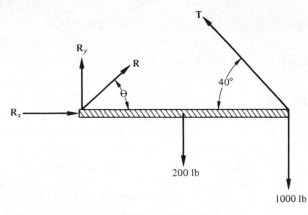

Figure 5-8.

rium, and taking moments about the left end of the boom, we can immediately obtain the tension T. Note that although we were not given the length of the boom, all distances can be expressed as some fraction of this length, so that the length appears on both sides of the moment equation and cancels.

Taking moments about the left end gives,

$$200 \times \frac{L}{2} + 1000L = T(\sin 40°) L = 0.6428 T \times L$$

$$1100 = 0.6428 T$$

$$T = 1711 \text{ lbs.}$$

We obtain R_x using $\Sigma F_x = 0$, which gives

$$R_x = T \cos 40° = 0.7660 \times 1711$$

$$R_x = 1311 \text{ lbs.}$$

The use of $\Sigma F_y = 0$, gives us R_y

$$R_y + T \sin 40° = 1000 + 200$$

$$R_y = 1200 - 0.6428 \times 1711$$

$$= 1200 - 1100 = 100 \text{ lbs.}$$

The force R exerted on the boom by the wall is given by

$$R = \sqrt{R_x^2 + R_y^2} = \sqrt{(1311)^2 + (100)^2} = 1314 \text{ lbs.}$$

The angle Θ, between R and the boom is given by:

$$\tan \Theta = \frac{R_y}{R_x} = \frac{100}{1311} = 0.0763 \quad \text{and} \quad \Theta = 4.33°.$$

Example 5-8. A ladder 40 ft long, weighing 80 lbs, rests against a smooth wall and makes an angle of 60° with respect to the ground. If the coefficient of friction between ladder and ground is 0.2, how far up the ladder will a 200 lb man be able to go before the ladder slips? What must the coefficient of friction be (minimum) if the man is to go to the top of the ladder without it slipping. Because the wall is smooth we neglect friction there and assume that R, the force exerted by the wall on the ladder, is perpendicular to the wall (see Figure 5-9).

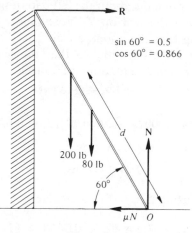

$$\sin 60° = 0.5$$
$$\cos 60° = 0.866$$

Figure 5-9.

Solution. Using equation 5.1, the first condition for equilibrium, we obtain both N and R.

$$\Sigma F_y = 0 \quad \text{gives} \quad 200 + 80 = N = 280 \text{ lbs}$$
$$\Sigma F_x = 0 \quad \text{gives} \quad \mu N = R = 0.2 \times 280 = 56 \text{ lbs.}$$

Using the second condition for equilibrium, and taking moments about the base of the ladder, at point O, we get:

$$200 \times d \cos \Theta + 80 \times (L/2) \times \cos \Theta = R \times L \times \sin \Theta.$$

Here d is the distance the 200 lb man goes up the ladder, and L is the length of the ladder. Inserting the appropriate values,

$$200 \times d \times 0.5 + 80 \times 20 \times 0.5 = 56 \times 40 \times 0.866 = 1940$$
$$d + 8.0 = 19.4$$

and

$$d = 11.4 \text{ ft.}$$

To determine the minimum value of μ which would permit the man to climb to the top of the ladder, we use the preceding moment equation, replacing d with L, the length of the ladder, and solve for μ, setting $\mu N = R$ as required by $\Sigma F_x = 0$.

$$200L \cos \Theta + 80 \times 20 \cos \Theta = RL \sin \Theta = NL \sin \Theta$$
$$200 \times 40 \times 0.5 + 80 \times 20 \times 0.5 = 280\mu \times 40 \times 0.866$$
$$40 + 8 = 97\mu$$
$$\mu = \frac{48}{97} = 0.495.$$

Chapter 6
Work and Energy

6-1. Work. Work is done on an object when a force acts on the object and causes it to move through some distance d. The amount of work done is defined as the product of the magnitude of the component of the force in the direction of the motion, and the magnitude of the displacement or the distance moved, so that

$$W = Fd \cos \Theta. \tag{6.1}$$

In the M.K.S. system, work has the units newton-meters, and 1 newton meter is called a joule. In the English gravitational system the natural unit is the foot pound, for which no special name exists.

This definition is somewhat more narrow than the everyday usage of the word "work," which is considered synonymous with "effort." This difference can be illustrated by a simple example. No one would argue with the statement that it is "a lot of work" to carry a heavy suitcase, for example, from Grand Central Station in New York to Penn Station. If we assume, however, that the trip is made along a level surface and that the suitcase is held at a *constant* height above the sidewalk, then no work is done, in a physical sense, except in originally lifting the suitcase to this height and getting it started. This is true because the force exerted on the suitcase is vertical and the displacement is horizontal and cos Θ in equation 6.1 is zero.

Again, no one would disagree that lifting the suitcase, walking a few steps, stopping and putting it down was a negligible part of the effort involved in this trip. The answer, of course, is that even standing in one place and holding a suitcase, we do what might be called physiological work in that our muscles are constantly flexing and changing by small amounts. It also would be very difficult to walk in a normal, somewhat bouncy manner and not raise and lower the suitcase by a slight amount. In any event, no net work is done *on the suitcase* unless its height is increased, or it is accelerated.

Note that work is a scalar quantity, since we have defined it in terms of the magnitudes of the force and displacement involved. There is no direction associated with work.

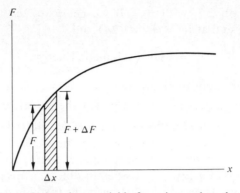

Figure 6-1. The work done by a variable force is equal to the area under the curve of force vs. distance.

Equation 6.1 assumes that the force is constant, but of course many situations occur in nature in which work is done on an object by a variable force. Figure 6-1 is a graph of force versus distance for such a situation. The nearly rectangular area of width Δx and height F is approximately equal to the work done as the force changes from F to $F + \Delta F$ and x changes from x to $x + \Delta x$. As Δx becomes smaller and smaller, this area becomes more and more nearly equal to $F\Delta x$. If we think of the area between the curve and the x axis being completely filled by a continuous series of such narrow rectangles (and let Δx become very small), it should be clear that the sum of the areas of these rectangles approaches the area under the curve. Therefore, this area must be equal to the work done as the variable force F acts through some distance x.

6-2. Kinetic Energy. The term energy is used to indicate the capacity to do work. Certainly an object which is moving has this capacity, since it can collide with another object and cause it to move and thus do work on the object. Let us calculate the amount of work required to accelerate an object from rest to a velocity v. We consider an object of mass m at rest on a frictionless, horizontal surface. A constant force of magnitude F parallel to the surface acts on the object and moves it at constant acceleration through a distance d, after which its velocity is v. Since $W = Fd$, and $F = ma$, we have:

$$W = Fd = mad.$$

From equation 2.7, with $v_0 = 0$ we get $v^2 = 2ad$, from which $a = v^2/(2d)$, so that $W = mv^2d/(2d)$, or

$$W = \tfrac{1}{2}mv^2 = \text{Kinetic Energy.} \qquad (6.2)$$

Thus, the work done in giving an object of mass m, a velocity v, is $\frac{1}{2}mv^2$, and this latter quantity is called the kinetic energy of the object. Before proceeding to examples which illustrate the usefulness of the energy concept, we discuss in the sections which follow another type of energy and the principle of energy conservation.

6-3. Potential Energy.   Consider an object of mass m which is raised vertically from some starting point through a height h, as indicated in Figure 6-2. A constant force equal to the force of

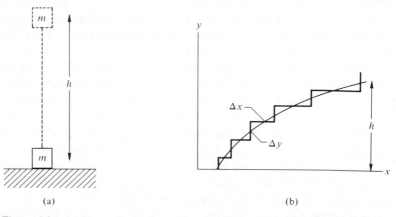

Figure 6-2.   (a) The work done in raising an object of mass m through height h is mgh. (b) The work done is independent of the path and depends only on the net change in the vertical height.

gravity, mg, must be exerted on the object merely to support it, and we assume it moves, with negligible acceleration, to a height h. The work done, according to equation 6.1, must be:

$$W = P.E. = mgh. \qquad (6.3)$$

Relative to its starting position, we say that the object now has a gravitational potential energy of mgh. By virtue of its position, it can fall through the distance h and do work. It should be apparent that in so doing its potential energy is converted into kinetic energy, and that after falling a distance h all its potential energy has been converted into kinetic energy, so that $\frac{1}{2}mv^2 =$ mgh, and $v = \sqrt{2gh}$. Figure 6-2(b) illustrates the fact that the work done on any object in being raised through a vertical height h, is independent of the path taken. By considering an arbitrary path and then approximating it with a series of steps in the x and y direction we see that only in the vertical steps, Δy, is work done, since the force due to gravity acts along the vertical. The sum of all the Δy's must add up to h, so that the work done as a result of the vertical steps depends only on h. We can take many, very small increments in x and y and approach any arbitrary path as closely as desired.

The value of the potential energy of an object at a particular location might seem rather arbitrary, and indeed it is. One may choose any reference point desired from which to calculate the potential energy of an object, as long as this point is sufficiently close to the surface of the earth that the force due to gravity, mg, does not change. The reason for this seemingly strange situation is that the only important quantity in any application of the potential energy concept is the *difference* in potential energy between two positions and this must always be the same no matter where we locate the zero of potential energy.

Another example of potential energy is the energy stored in a compressed or extended coil spring. For most springs, the relationship between the applied force and the amount of compression or extension is $F = kx$, where F is the force, x is the compression or extension from the equilibrium position, and k is the constant of proportionality or force constant. This linear relationship, called Hooke's law, is found to hold for values of x which are small enough so that the spring is not stretched beyond its so-called elastic limit and does not return to its original shape. Although the force applied to a spring to compress it (for example, an amount x) is not constant, it is easy to calculate the average force because of the linear relationship $F = kx$. The force starts at zero (Figure 6-3(a)), and at a compression of x (Figure 6-3(b)) becomes equal to kx. Thus, the average force becomes $(0 + kx)/2 = \frac{1}{2}kx$. The work done in compressing the spring (or in stretching it) an amount x is simply:

$$W = F_{av}x = \left(\frac{1}{2}kx\right)x = \frac{1}{2}kx^2. \tag{6.4}$$

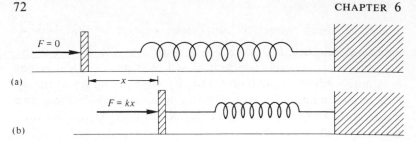

Figure 6-3. The elastic potential energy stored in a compressed spring is $\frac{1}{2}kx^2$.

The quantity $\frac{1}{2}kx^2$ is known as the elastic potential energy stored in the spring. The same result is obtained if the spring is stretched a distance x.

6-4. Conservation of Energy. We have already mentioned the fact that the potential energy of an object is converted into kinetic energy as it falls under the influence of gravity. Another example of the conversion of potential energy to kinetic energy and vice-versa is illustrated by the simple pendulum in Figure 6-4. When the pendulum reaches its maximum height its velocity is zero, its kinetic energy is zero, and its potential energy is a maximum. At the lowest point in its swing the kinetic energy is a maximum and equal to the *change* in potential energy. The total energy of the system at any point is equal to the sum of the kinetic energy and the potential energy, and for the pendulum this is clearly equal to *mgh*, corresponding to maximum potential energy and zero kinetic energy.

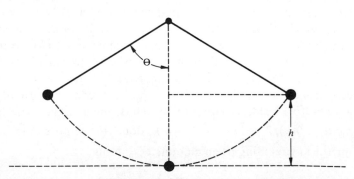

Figure 6-4. For a simple pendulum, neglecting frictional losses, the total energy—kinetic plus potential—is a constant.

For many systems in nature, such as the simple pendulum, the total energy is apparently constant at least for a reasonable length of time, and we say that the total energy is conserved. We know, however, that the pendulum will not swing back and forth forever, and that its amplitude will slowly decrease because of friction at the point of support and because of air resistance. If frictional losses are small, it is possible to use the principle of conservation of total energy as the basis for the solution of many problems in the field of mechanics. This principle of mechanical energy conservation is part of a general conservation principle known as *conservation of energy*, which, to the best of our knowledge, applies to any system we might wish to consider. This general principle of energy conservation applies to all types of energy, though so far we have discussed only kinetic and potential energy for mechanical systems. It states that energy is neither created nor destroyed but merely changes from one form to another. Probably the most familiar energy transformation is from kinetic energy to heat energy. This occurs whenever friction is encountered. Rubbing our hands together makes them warm; the brakes on an automobile or train will overheat if used excessively; and a squash ball warms up after repeated vigorous collisions with the front wall of the court. The examples of this type are numerous. Experiments have indicated that in situations such as those mentioned above, all the kinetic energy goes into heat energy. The reverse, however, is not true, and it turns out to be impossible to convert a given amount of heat energy into mechanical energy without using additional energy from some other source. This is known as the second law of thermodynamics.

It is interesting to note that thermal energy in matter is really kinetic and potential energy in a slightly different form. For example, when solids are heated, the additional energy goes into increasing the amplitude of vibration of the atoms about their equilibrium positions in the solid, while in gases it is found that the average kinetic energy of the molecules is directly proportional to the absolute temperature.

We will now apply the energy principle to several examples to illustrate its wide range of usefulness. Note that some of these examples require that friction be neglected, while others are directly concerned with the amount of energy dissipated as heat and require no such approximation.

Example 6-1.   A ball is dropped from a height of 25 ft. Using conservation of energy, calculate the velocity of the ball as it hits the ground.

Solution. We equate the kinetic energy as it hits the ground to the *change* in potential energy and obtain:

$$\tfrac{1}{2}mv^2 = mgh, \quad \text{from which}$$
$$v = \sqrt{2gh} = \sqrt{2 \times 32 \times 25} = 40 \text{ ft/sec.}$$

Note that this is the same result as would be obtained using equation 2.7 $(v^2 = v_0^2 + 2ad)$.

Example 6-2. A wooden block is given a velocity of 4 meters per second across a horizontal surface. The kinetic coefficient of friction between the block and the surface is 0.4. How far will the block slide?

Solution. We apply energy conservation and note that the block will continue to move until all its kinetic energy has been converted to heat energy. This gives us:

$$\tfrac{1}{2}mv^2 = F_f x = \mu N x.$$

Here $N = mg$ and F_f = force of friction, and

$$x = \frac{\tfrac{1}{2}mv^2}{(\mu N)} = \frac{0.5v^2}{(\mu g)}$$
$$x = \frac{0.5 \times (4)^2}{(0.4 \times 9.8)} = 2.04 \text{ meters.}$$

Example 6-3. (a) From what height above the bottom of the loop must the car in Figure 6-5 start in order to "just make it" around the loop? (b) What is the velocity of the car at point A and at point B?

Solution. (a) If the car is to "just make it" around the loop its speed must be such that the force of gravity on it is sufficient to provide the centripetal force. For this to be the case, $mv^2/r = mg$ or $v = \sqrt{rg}$. Neglecting friction, we use conservation of energy and note that the velocity of the car at point D must be the same as at point C, since both correspond to the same net change in potential energy relative to the starting point. The velocity at point D is obtained by setting the kinetic energy equal to the change in potential energy, giving $\tfrac{1}{2}mv^2 = mgh$, where h is the vertical height of the starting point above points C and D.

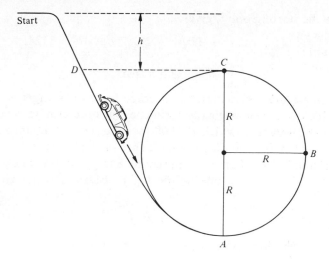

Figure 6-5.

Thus $v = \sqrt{2gh}$, and since v should be equal to $\sqrt{rg}$ at the top of the loop, we get $2gh = rg$, which gives $h = r/2$, indicating that the starting point must be $2r + r/2 = \frac{5}{2}r$ above the bottom of the loop. (b) The speed at point A is easily obtained using the results of part (a). We know that v must be given by $\sqrt{2gh}$, and h in this case is $\frac{5}{2}r$, so that $v = \sqrt{5gr}$.

At point B the velocity is obtained by proceeding as above, using h as $\frac{3}{2}r$, since h is the net change in the vertical position. This gives $v = \sqrt{3rg}$.

Example 6-4. A car moves at a constant speed of 10 ft/sec along a level road. It then rolls down a steep hill to another level stretch which is 77 ft lower than the first. Finally, it rolls up another hill to a third level stretch of road which is 61 ft above the lowest portion of the road. What is the speed of the car on the two lower level stretches of the road. Neglect friction, and assume that the motor is off.

Solution. If $\frac{1}{2}mv_0^2$ is the original kinetic energy, the kinetic energy on the lowest part of the road must be $\frac{1}{2}mv^2 = \frac{1}{2}mv_0^2 + mgh$, where h is 77 ft. This means that $v^2 = v_0^2 + 2gh = (10)^2 + 2 \times 32 \times 77 = 100 + 4928 = 5028$, and $v = 70.9$ ft/sec. The same procedure applied to the third stretch of road, which is 16 ft below the starting point, gives:

$$v^2 = v_0^2 + 2gh = (10)^2 + 2 \times 32 \times 16 = 1124,$$

and

$$v = 33.5 \text{ ft/sec}.$$

Example 6-5. Calculate the average force on a large, wooden log which is driven vertically six inches into the ground by a pile driver the weight of which is 2000 lbs, and which falls through a distance of 10 ft (including the six inches the log moves).

Solution. The change in potential energy of the pile driver must equal the work done as the log is driven into the ground, i.e.,

$$mgh = F_{av}d,$$

where d is one half foot. Thus,

$$F_{av} = \frac{mgh}{d} = 2000 \times \frac{10}{0.5} = 40,000 \text{ lbs}.$$

Example 6-6. A large coil spring is placed at the rear of a garage and must be designed to permit beginning drivers of a 3200 lb car to hit the rear of the garage at 15 mph and only compress the spring 2 ft. What must the spring constant k of this spring be?

Solution. The kinetic energy of the car at 15 mph (22 ft/sec) must all be stored as elastic potential energy in the spring. Therefore:

$$\tfrac{1}{2}mv^2 = \tfrac{1}{2}kx^2,$$

and

$$k = \frac{mv^2}{x^2} = \frac{100 \times (22)^2}{(2)^2} = 1.21 \times 10^4 \text{ lbs/ft}.$$

Example 6-7. The amount of heat energy required to raise the temperature of 1 gram of a substance by one degree centigrade is called the specific heat. Heat energy is often measured in units called calories. 1 calorie = 4.186 joules. Thus, the heat required to raise the temperature of M grams of a material with a specific heat C an amount ΔT is, $H = MC\Delta T$. If a rapidly moving 100 g projectile ($v = 300$ m/sec) is stopped inside a 2 kg block of aluminum which is fixed so that it cannot move, calculate the temperature rise of the aluminum, assuming that all the kinetic energy of the projectile is converted into heat energy in the block. Neglect the fact that the bullet, of mass m, also must be heated up. The specific heat of aluminum is 0.2 cal/g°C.

Solution. The amount of kinetic energy available is:

$$\tfrac{1}{2}mv^2 = 0.5 \times 0.1 \times (300)^2 = 4500 \text{ joules} \approx 1075 \text{ cal.}$$

If all the heat goes to 2 kg of aluminum,

$$H = MC\Delta T,$$

and

$$\Delta T = \frac{H}{(MC)} = \frac{1075}{(2000 \times 0.2)}$$

$$\Delta T = 2.69°C.$$

6-5. Power. Often it is important to specify not only how much work is done, but the length of time in which it is done. Certainly one measure of the performance of any kind of a motor is the rate at which it will do work. This rate is called the power and is defined as

$$P = \frac{\Delta W}{\Delta t} \tag{6.5}$$

where ΔW is the work done in the time interval Δt. In the M.K.S. system the unit of power is *the joule/sec* or the *watt*. In the English gravitational system the unit of power is the ft lb/sec, and the familiar horsepower is defined as 550 ft lbs/sec, which is equivalent to 746 watts.

Although the watt may be familiar as a unit of electric power, note that there is nothing basically electrical about its definition. Electric power companies bill their customers in terms of a unit called the kilowatt hour, which is the amount of work done in one hour by a device which uses energy at the rate of one kilowatt (10^3 watts). Thus the kilowatt hour is a unit of energy and is equal to 3.6×10^6 joules.

Let us consider an object which is acted on by a force F. The object moves through a distance Δx and an amount of work $\Delta W = F\Delta x$ is done. The rate at which work is done, or the power, is given by $\Delta W/\Delta t = F\Delta x/\Delta t$. Since $\Delta x/\Delta t = v$, we obtain:

$$P = Fv \tag{6.6}$$

which is a useful relationship between the instantaneous power, force, and velocity obtained by making Δt very small, so that the rates $\Delta x/\Delta t$ and $\Delta W/\Delta t$ become the instantaneous velocity, and power, respectively.

Example 6-8. A 200 lb man carrying a 20 lb pack climbs from the valley to the top of a mountain in 2 hours. If the vertical height through which he moved was 2000 ft, calculate the average rate at which he did work in making the climb.

Solution. The work required to raise 220 lbs through a distance of 2000 ft is $W = mgh = 220 \times 2000 = 440,000$ ft lbs. The average rate of doing work or the average power is $\Delta W/\Delta t = 4.4 \times 10^5/7200 = 61.1$ ft lbs/sec. Since 1 horsepower equals 550 ft lbs/sec, the average power is $61.1/550 = 0.111$ hp. Note that 1 kilogram calorie (a dietician's calorie) equals 3100 ft lbs, so that in terms of the physical work only about 142 calories were used up. This number is misleading because the climber used many times this amount of energy physiologically, but the above calculation gives some idea of the difficulty involved in losing large amounts of weight by exercising over a relatively short period.

Example 6-9. A 3200 lb car is accelerating at the rate of 10 ft/sec². Calculate the power delivered by the engine to the car at the instant the car has a velocity of 50 ft/sec.

Solution. The force exerted by the car on the road must be given by $f = ma = \frac{3200 \times 10}{32} = 1000$ lbs. Therefore, the power, according to equation 6.6 must be $P = Fv = 1000 \times 50 = 50,000$ ft lbs/sec or 90.9 hp.

6-6. Rotational Kinetic Energy. Clearly, a rotating object, even if its center of mass is stationary, must have kinetic energy because all parts of it are in motion. Let us consider a rotating object, such as a grinding wheel or a circular saw blade, and obtain an expression for its rotational kinetic energy. In Figure 6-6 a circular disc is shown rotating about an axis through its center with an angular velocity ω. Consider a very small part of the disc of mass m, located a distance r from the axis of rotation. This particle is moving with a velocity $v = \omega r$ and its kinetic energy must be $\frac{1}{2}mv^2 = \frac{1}{2}m\omega^2 r^2$. If we examine all of the particles which make up the disc and add their kinetic energies, we obtain for the rotational kinetic energy of the disc:

$$(K.E.)_{rot} = \tfrac{1}{2}\omega^2(m_1 r_1^2 + m_2 r_2^2 + m_3 r_3^2 + \cdots).$$

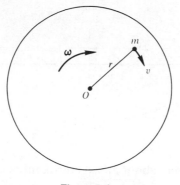

Figure 6-6.

The quantity

$$I = (m_1 r_1^2 + m_2 r_2^2 + \cdots) \tag{6.7}$$

is called the moment of inertia of the rotating object and is usually designated by the symbol I. As indicated from equation 6.8, the moment of inertia for rotational motion

$$K.E. = \tfrac{1}{2} I \omega^2 \tag{6.8}$$

is analogous to the mass in the expression for linear or translational kinetic energy. The moment of inertia depends upon the size, shape, and density of the rotating object and, of course, upon the location of the axis of rotation. For irregular shaped objects it is virtually impossible to calculate the moment of inertia about a particular axis, but for symmetrical objects—such as circular discs, spheres or uniform rods—the moment of inertia can be calculated in a straightforward manner using the calculus. For a uniform, circular disc, rotating about an axis, through its center, the moment of inertia is given by $I = \tfrac{1}{2} m r^2$ where m is the mass of the disc and r is its radius. For a solid sphere, rotating about its center, the moment of inertia is $\tfrac{2}{5} m r^2$, while for a thin, spherical shell the moment of inertia about the same axis is $\tfrac{2}{3} m r^2$. The units of the moment of inertia in the M.K.S. system are kg m^2.

Example 6-10. A rigid rod of negligible mass supports four, 5 kg masses as indicated in Figure 6-7. Calculate the moment of

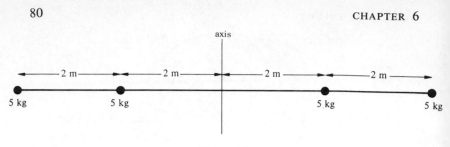

Figure 6-7.

inertia of this system about an axis through its center perpendicular to the rod.

Solution. To calculate the moment of inertia we must calculate the sum of the products mr^2, as indicated in equation 6.7. Here our problem is simplified because we have only four masses, all of which are located at a definite distance from the axis. For the purposes of the problem we assume these masses to be point masses, i.e., that their size is small compared to 2 meters. The product mr^2 for each of the two inner masses is $5 \times (2)^2 = 20$ kg m^2, while mr^2 for each of the two outer masses is $5 \times (4)^2 = 80$ kg m^2. Thus the two inner masses together contribute 40 kg m^2, while the two outer masses contribute 160 kg m^2, making a total of $40 + 160 = 200$ kg m^2, which gives the moment of inertia for this system. Note the substantially larger contribution made by the outer masses because of the dependence of the moment of inertia on the individual values of r^2.

Example 6-11. A large cylindrical grinding stone ($I = \frac{1}{2}mr^2$) of radius 50 cm, and mass 60 kg, rotates about its center with an angular velocity of 20 radians per second. If it is assumed that sharpening an axe at the rim of the wheel exerts a constant force of 20 newtons, calculate how many turns the wheel will make before it stops.

Solution. The rotational kinetic energy of the rotating grinding wheel is gradually converted into heat as the wheel does work on the axe. We can equate the original rotational kinetic energy of the wheel to the product Fd where F is 20 newtons, and d is the length measured along the rim of the wheel which passes under the axe blade, and Fd is the work done on the axe. Thus:

$$\tfrac{1}{2}I\omega^2 = Fd, \quad \text{and} \quad d = \frac{\tfrac{1}{2}I\omega^2}{F}$$

$$d = \tfrac{1}{2} \times \tfrac{1}{2} \times 60 \times 0.5^2 \times \frac{20^2}{20} = 75 \text{ meters.}$$

The circumference of the wheel is $2\pi r = 2 \times 3.14 \times 0.5 = 3.14$ m. Therefore, the wheel must turn through a number of revolutions given by d divided by the circumference or $75/3.14 = 23.9$ revolutions.

When an object such as a disc or a sphere is rolling, it has kinetic energy due to its linear velocity as well as kinetic energy due to its rotation. Its total kinetic energy is the sum of the translational and rotational kinetic energies.

Example 6-12. Consider a circular disc and a thin-walled cylinder. Each rolls, without slipping, down an inclined plane, and in so doing moves through a vertical height h. For each, calculate the fraction of its kinetic energy which appears as translational kinetic energy when it reaches the bottom.

Solution. We use conservation of energy and in each case equate the change in potential energy to the gain in kinetic energy. For the solid, circular disc, the moment of inertia is $\tfrac{1}{2}mr^2$, and we obtain:

$$mgh = \tfrac{1}{2}mv^2 + \tfrac{1}{2}I\omega^2,$$

and since $v = \omega r$,

$$mgh = \tfrac{1}{2}mv^2 + \tfrac{1}{2} \times \tfrac{1}{2}mr^2 \times \frac{v^2}{r^2}$$

$$mgh = \tfrac{1}{2}mv^2 + \tfrac{1}{4}mv^2$$

so that only $\tfrac{2}{3}$ of the kinetic energy appears as translational energy, and $\tfrac{1}{3}$ as rotational kinetic energy.

For the thin-walled cylinder, essentially all of the mass is at a distance r from the center, and its moment of inertia about an axis through the center is mr^2.

$$mgh = \tfrac{1}{2}mv^2 + \tfrac{1}{2}I\omega^2$$

$$mgh = \tfrac{1}{2}mv^2 + \tfrac{1}{2}mr^2 \times \frac{v^2}{r^2} = \tfrac{1}{2}mv^2 + \tfrac{1}{2}mv^2$$

and the energy is evenly divided between rotational and translational kinetic energy.

Chapter 7

Momentum

7-1. Introduction. Momentum is the name given to the quantity defined as the product of the mass and velocity of an object. It is a vector quantity, having both magnitude and direction, its direction being the same as that of the velocity.

$$\mathbf{p} = m\mathbf{v}. \tag{7.1}$$

As indicated in equation 7.1, it is customary to represent the momentum with the letter $\mathbf{p}$. The units of momentum in the M.K.S. system are kg m/sec, and in the English gravitational system, slug ft/sec. Momentum is a particularly useful concept in dealing with problems involving collisions and it is fully as important and useful as the concept of energy. During the last decade of the 17th century and for about forty years of the 18th century, a vigorous controversy developed between the famous scientists Descartes and Leibnitz and their followers as to whether the proper way to measure the effect of a force acting on an object was in terms of the resulting kinetic energy or momentum. It turned out that both were correct and that they were talking about somewhat different concepts. As we shall see below, the change in momentum experienced by an object acted on by a constant force is simply the product of the force and the *time*, while the kinetic energy change is given by the product of the force and the *distance*. Scientists soon realized that both kinetic energy and momentum are important fundamental concepts and that neither should be singled out. We use either or both, depending upon the nature of the problem.

It is interesting to note that Newton originally expressed his second law in terms of the momentum concept. In this form

$$\mathbf{F} = \frac{\Delta \mathbf{p}}{\Delta t} \tag{7.2}$$

the second law says that the vector sum of the forces on a body is equal to the rate of change of momentum. This is expressed in equation 7.2 where $\Delta \mathbf{p}$ is the change in momentum experienced in the time Δt. It is easy to see that if the mass m is constant, this expression is equivalent to equation 4.1. Since $\mathbf{p} = m\mathbf{v}$ we can write $\mathbf{F} = \dfrac{\Delta(m\mathbf{v})}{\Delta t} = \dfrac{m\Delta\mathbf{v}}{\Delta t}$, and since $\mathbf{a} = \Delta\mathbf{v}/\Delta t$, equation 7.2 reduces to the familiar $\mathbf{F} = m\mathbf{a}$, if m is constant. Equation 7.2 is actually the more general expression. It turns out that the mass of an object depends on its velocity, as predicted by the special theory of relativity, and if this velocity is nearly equal to the velocity of light, the mass can be thousands of times larger than the mass at low speeds. This is particularly important for subatomic particles such as the electron and proton which can easily be given speeds close to that of light.

7-2. Conservation of Linear Momentum. If we express the second law as $\mathbf{F} = \Delta(m\mathbf{v})/\Delta t$, and note that $\Delta(m\mathbf{v}) = \mathbf{F}\,\Delta t$, it is clear that unless some force acts on the object under consideration, there can be no change in its momentum. When this fact is applied to a system of two or more objects it becomes clear why the momentum concept is so useful.

Suppose that we have a system made up of many objects and that no unbalanced forces are exerted on the objects by the surroundings. A collection of billiard balls on a level, frictionless surface might be such a system. The total momentum of the system must be the vector sum of the momenta of the various objects. When the objects collide and experience what we might call "internal forces," i.e., forces within the system, the momentum changes resulting from these forces must cancel out in pairs, because by Newton's third law each object experiences a force equal and opposite to that experienced by the object with which it collides. Thus, the resulting momentum changes must be equal and opposite. This reasoning leads us to the law of conservation of linear momentum, according to which the total linear momentum of a system remains constant if the vector sum of all the external forces acting on the system is zero. Conservation of momentum ranks with energy conservation as one of the most fundamental laws of nature. As an example of the type of problem to which momentum conservation may be applied, consider the following:

Example 7-1. The two blocks shown in Figure 7-1 move in opposite directions along the same line on a horizontal frictionless surface. When they collide, they stick together. After the collision, in what direction and with what speed do the combined blocks move?

Solution. We consider a "system" composed of the two

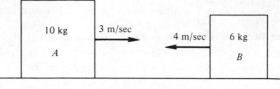

Figure 7-1.

blocks. Since the momentum of block A is 30 kg m/sec to the right, and that of block B is 24 kg m/sec to the left, the net momentum of the system is the vector sum of the two momenta, or 6 kg m/sec to the right. Conservation of momentum requires that after the collision the momentum of the system, now the combined blocks, must be the same.

Since the mass of the two blocks stuck together is 16 kg, and since the momentum of this combination must be 6 kg m/sec to the right we get: $(M_A + M_B)v = 6$, or $16v = 6$, so that, $v = \frac{6}{16} = 0.375$ m/sec to the right.

In the above example, the collision was completely inelastic. If the blocks had collided and bounced apart so that kinetic energy was conserved, i.e., so that the sum of the kinetic energies was the same before and after the collision, the collision would be described as elastic. All collisions fall somewhere between these extremes, and although momentum must be conserved in *all* collisions, this fact alone is not sufficient to determine the motion of the objects after the collision. The degree of elasticity must somehow be specified. One way of doing this is to specify the velocity of one of the objects after the collision. If this information is available, the velocity of the other is uniquely determined by momentum conservation.

Example 7-2. As shown in Figure 7-2, a 4 kg ball, sliding with a velocity of 5 m/sec on a smooth surface with negligible friction, collides with a stationary 3 kg ball. The 3 kg ball moves off with a velocity of 4 m/sec at an angle of 30° with respect to the direction of the incident ball. At what angle and with what velocity must the incident ball move after the collision? Was energy conserved in the collision?

Solution. The geometry of the collision is indicated in Figure 7-2(a).

We make specific use of the vector nature of momentum as indicated in Figure 7-2(b), which applies to the 3 kg ball; and in

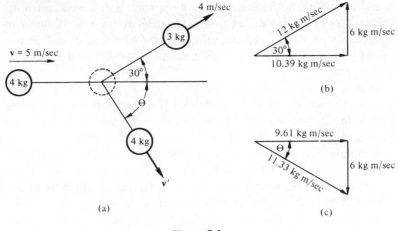

Figure 7-2.

Figure 7-2(c) which applies to the 4 kg ball. We are given the momentum of the 3 kg ball after the collision. Since the x and y components of the "system," i.e., both balls, are the same before and after the collision, the y components of the two balls must be equal and *opposite* after the collision. The y component of the 3 kg ball is $12 \times \sin 30° = 12 \times 0.5 = 6$ kg m/sec. Thus, the y component of the 4 kg ball must also be 6 kg m/sec, as indicated in (c). Similarly the x components of the momenta of the two balls must add to 20 kg m/sec. For the 3 kg ball, $12 \times \cos 30° = 10.39$ kg m/sec. Thus, for the 4 kg ball the x component of the momentum must be $20 - 10.39 = 9.61$ kg m/sec. Knowing the x and y components of the 4 kg ball, we use the Pythagorean theorem to calculate $p = \sqrt{(9.61)^2 + (6)^2} = 11.33$ kg m/sec. Since $p = mv$, we get $v = 11.33/4 = 2.83$ m/sec. From the vector triangle of (c) we get $\tan \Theta = 6/9.61 = 0.624$ and $\Theta = 32°$. The kinetic energy before the collision is $K.E. = \frac{1}{2}mv^2 = \frac{1}{2} \times 4 \times (5)^2 = 50$ joules. After the collision, for the 4 kg ball, $\frac{1}{2}mv^2 = \frac{1}{2} \times 4 \times (2.83)^2 = 16.04$ joules. For the 3 kg ball, $\frac{1}{2}mv^2 = \frac{1}{2} \times 3 \times (4)^2 = 24$ joules. Thus, the kinetic energy after the collision is $24 + 16 = 40$ joules, whereas before the collision the total kinetic energy was 50 joules. The collision was partially inelastic, and kinetic energy was not conserved.

Example 7-3. A spring is compressed between two toy railroad cars. With both cars at rest the spring is released, pushing the cars away from each other in opposite directions. If one car has a mass of 10 kg, and the other a mass of 7 kg, calculate the velocity of the 7 kg car if the 10 kg car is found to have a velocity of 2 m/sec.

Solution. Since the momentum of both is zero before the release of the spring, the vector sum of the momenta must also be zero after it is released. Therefore, we can say that $M_{10}v_{10} = M_7v_7$ and $v_7 = v_{10} \times M_{10}/M_7 = 2 \times 10/7 = 2.86$ m/sec.

Example 7-4. A baseball traveling with a speed of 80 ft/sec parallel to the ground is hit by a bat and leaves the bat in the opposite direction with a speed of 100 ft/sec. If the ball is assumed to have a weight of 1/3 lb, and to have been in contact with the bat for 0.005 sec, calculate the average force exerted by the bat on the ball.

Solution. Here we utilize the fact that $\Delta\mathbf{p} = \mathbf{F}_{av} \Delta t$. The momentum of the ball before striking the bat is $(0.33/32) \times 80 = 0.0103 \times 80 = 0.825$ slug ft/sec. After the collision it is $0.0103 \times 100 = 1.03$ slug ft/sec in the opposite direction. Thus, the *change* in momentum is $0.825 + 1.03 = 1.86$ slug ft/sec. Momentum is a vector quantity and Δp is the magnitude of the vector which, when added in the opposite direction to the original momentum, gives the momentum after the collision. Thus, $\mathbf{F}_{av} = \Delta\mathbf{p}/\Delta t = 1.86/0.005 = 372$ lbs.

Example 7-5. A machine gun shoots bullets into a large wooden board at the rate of 10 per second. Each bullet has a velocity of 400 m/sec and a mass of 5 grams. The bullets penetrate the wood and stop and do not bounce off. What is the average force exerted by the bullets on the board?

Solution. The average force exerted by the bullets is given by the rate of change of momentum which they experience. Since the bullets do not bounce off the board, each bullet delivers a momentum $mv = 0.005 \times 400 = 2$ kg m/sec. Since 10 bullets hit the board each second, the rate of change of momentum per second must be $10 \times 2 = 20$ kg m/sec. Thus $\Delta p/\Delta t = 20$ kg m/sec^2 = 20 newtons.

Example 7-6. Suppose that the machine gun of Example 7-5 fired hard steel spheres of mass 5 grams, and that these spheres bounced off a steel plate with the same speed (400 m/sec) at which they hit the plate and in the opposite direction. What is the average force exerted by the balls on the steel plate?

Solution. Our answer must be 40 newtons, which is twice the value obtained in Example 7-5. The reason for this is simply that the momentum change experienced by each ball is *twice* what it would be if the ball stopped in the plate. Therefore, the rate of change of momentum experienced by the plate, which equals the force experienced by the plate, must also be twice as large.

Figure 7-3 illustrates a simple device known as a ballistic pendulum which can be used to measure the muzzle velocity of a

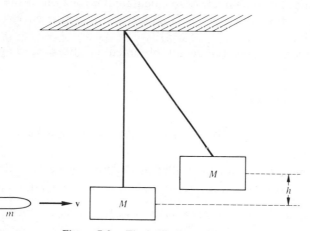

Figure 7-3. The ballistic pendulum.

bullet. The bullet is fired horizontally into the "bob" of the pendulum in which it is assumed to stop. Momentum conservation tells us that immediately after the collision, the bullet plus the "bob" moves to the right such that $mv = (m + M)V$ where m is the mass of the bullet, M is the mass of the "bob," v the velocity of the bullet, and V the velocity of "bob" plus bullet after the collision. We wish to determine the original velocity of the bullet which is given by $v = V(M + m)/m$. This relationship has been determined by the use of momentum conservation, but to determine V, we must use conservation of mechanical energy. During the actual collision, mechanical energy is certainly not conserved (see Example 7-7), but after the collision, as the bob and bullet swing up, mechanical energy is conserved, assuming that friction in the air and at the suspension point are negligible. When the bob ultimately reaches a maximum height h, the change in its kinetic energy must equal the change in potential energy.

In this case the change in potential energy is equal to the kinetic energy of the bob plus bullet immediately after the collision. Thus, $(M + m)gh = \frac{1}{2}(M + m)V^2$, so that $V = \sqrt{2gh}$. When this result is inserted in the expression for v above, we get $v = \sqrt{2gh}(M + m)/m$.

Example 7-7. A rifle bullet having a mass of 0.05 kg is fired into a pendulum "bob" having a mass of 10 kg. If the "bob" rises a vertical distance of 10 cm after the collision, calculate the velocity of the bullet. Also calculate the fraction of the kinetic energy of the bullet which is converted into heat energy in the collision.

Solution. Using the result obtained in the preceding paragraph we get:

$$v = \sqrt{2gh}\,\frac{(M + m)}{m}$$

$$v = \sqrt{2 \times 9.8 \times 0.1} \times \frac{10.05}{0.05} = 281.4 \text{ m/sec.}$$

The kinetic energy of the bullet before hitting the pendulum was $\frac{1}{2}mv^2 = \frac{1}{2} \times 0.05 \times (281.4)^2 = 1979.6$ joules. After the collision, the kinetic energy is $\frac{1}{2}(M + m)V^2$ or $(M + m)gh$, which is calculated more easily. Thus $(M + m)gh = 10.05 \times 9.8 \times 0.1 = 9.849$ joules and $1969.8/1979.6$ or 99.5% of the original kinetic energy of the bullet is converted into heat.

7-3. Angular Momentum. The rotational analog of linear momentum is angular momentum. We approach the definition of angular momentum for an extended object rotating about some axis by first considering a very simple situation. Consider a small object, such as a ball of mass m on the end of a string, moving in a horizontal circle of radius r. The angular momentum of the ball is defined as mvr, where v is the speed of the ball. In terms of the angular velocity $\boldsymbol{\omega}$, the angular momentum is $mvr = mr^2\boldsymbol{\omega}$ since $v = \boldsymbol{\omega}r$. Angular momentum is a vector quantity and is directed along the axis of rotation. This direction is determined by a right hand rule such that if the fingers of the right hand curl in the direction of rotation, the thumb, if extended, points in the direction of the angular momentum. Thus a particle rotating in the plane of this page in a counterclockwise sense would have an angular momentum directed up out of the page and perpendicular to it. To calculate the angular momentum of an extended object such as the rotating disc of Figure 6-6 (page 79), we simply apply the above definition to each particle in the disc. Each particle moves in a circular path about the axis of rotation and we compute the sum of the products $mr^2\omega$, these products all being vectors in the same direction. We get for this sum, $\omega(m_1 r_1^2 + m_2 r_2^2 + m_3 r_3^2 + \cdots) = I\omega$, where I is the moment of inertia. Note the similarity between the expression mv for linear momentum and $I\omega$ for angular momentum.

There is also a fundamental conservation law associated with angular momentum. The law of conservation of angular momentum states that if there is no net torque acting on a system, the angular momentum of the system must stay constant. As was the case with linear momentum, this law has been found to hold at all levels in nature, from the tiny nuclei of atoms to the massive planets moving about the sun in our solar system.

We see many examples of conservation of angular momentum in everyday life. Perhaps the most familiar is the figure skater who starts to spin with both arms and one of her legs extended. As she pulls in her extended leg and her arms so that they are close to the axis of rotation, she starts to spin much more rapidly. The explanation of this fact is that by pulling in her leg and arms she has changed her moment of inertia about the axis of rotation. The total mass is the same but more of it has been brought in closer to the axis. If the angular momentum $I\omega$ is to stay constant, the angular velocity must increase if the moment of inertia decreases. A similar effect can be seen when a diver leaves the board fully extended. He acquires a certain angular momentum when he leaves the board, and this stays constant, but when he goes into the tuck position to do a somersault in the air before entering the water, his moment of inertia is markedly decreased and his angular velocity is correspondingly increased. The process is reversed just before he hits the water if he comes out of the tuck position and enters the water head first fully extended.

Example 7-8. A large circular disc of mass 50 kg rotates on nearly frictionless bearings with an angular velocity of 10 radians per second about a vertical axis. A second disc of mass 60 kg, originally at rest, is set directly on top of the first. If both discs have a radius of 30 cm, calculate the resulting angular velocity of the two discs (see Figure 7-4).

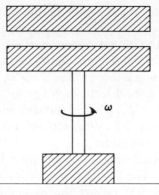

Figure 7-4.

Solution. The moment of inertia of a circular disc is $\frac{1}{2} Mr^2$, and if angular momentum is conserved, the angular momentum of the combined discs must be the same as the original angular momentum of the lower disc. Thus:

$$I_{50}\omega_0 = I_{50}\omega + I_{60}\omega$$

$$\omega = \frac{I_{50}\omega_0}{I_{50} + I_{60}} = \frac{\frac{1}{2} \times 50 \times (0.3)^2 \times 10}{\frac{1}{2} \times 50 \times (0.3)^2 + \frac{1}{2} \times 60 \times (0.3)^2}$$

$$\omega = \frac{50}{50 + 60} \times 10 = 4.55 \text{ rad/sec.}$$

Chapter 8

Universal Gravitation, Planetary Motion and Satellites

8-1. Newton's Law of Universal Gravitation. We have repeatedly used the fact that objects near the surface of the earth experience a gravitational force which results in an acceleration of about 32 ft/sec² for all freely falling objects. It was not until the 17th century, however, that scientists realized that this fact might result from an attraction between the earth and such objects. Sir Isaac Newton proposed a theory of universal gravitation in 1687 in his *Principia*, which explained all known gravitational phenomena, including the motions of the planets about the sun. Newton suggested that all objects attract one another with a force which is directly proportional to the product of the masses and inversely proportional to the square of the distance between them. Mathematically this is expressed as:

$$F = \frac{Gm_1 m_2}{r^2} \tag{8.1}$$

where m_1 and m_2 are the masses of two objects separated by a distance r. G is a constant of proportionality known as the universal gravitational constant, and is found experimentally to have the value 6.67×10^{-11} newton m²/kg². Such a constant of proportionality is necessary because we have already defined units of force, mass, and distance, and therefore cannot use this equation as we used equation 4.1 ($F = ma$) to define force or mass and conveniently make the proportionality constant equal to one.

Certainly one of the most impressive results of Newton's theory of gravitation was its complete explanation of the motion of the planets about the sun. Newton showed that the gravitational force of attraction between the sun and the planets provided the centripetal force required to keep them in their nearly circular orbits about the sun. Using equation 8.1, he was able to derive the experimental laws which the German mathematician, Johannes Kepler, had worked many years to discover. Kepler had found that the motion of the planets could be understood if one assumed that they moved in ellipses with the sun at one focus.

Kepler's three laws evolved from this assumption. His first law states that the planets move about the sun in ellipses with the sun at one focus. Kepler's second law says that as these planets move about the sun, a line drawn from the sun to the planet sweeps out equal areas in equal times as shown in Figure 8-1.

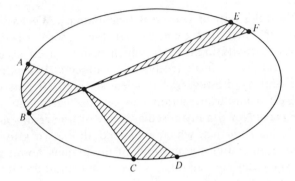

Figure 8-1. A planet moves in an elliptical orbit about the sun and the line joining them sweeps out equal areas in equal times.

The planet moves between points A and B in the same length of time required to move between C and D, and E and F, and the areas, ABO, CDO, and EFO are equal. The third law gives a relationship between the period of the planet and its average distance from the sun. The period is the time required for one revolution about the sun. The third law states that $T^2 = kr^3$, where T is the period, r is the mean radius of the orbit, and k is a constant of proportionality. These laws, the result of Kepler's analysis of the accurate observations of Tycho Brahe, the Danish astronomer, were the experimental facts which could not be explained or predicted by theory until Newton proposed his theory of universal gravitation.

Example 8-1. Show for the case of a circular orbit, that if a planet obeys Kepler's third law, it must be under the influence of a force which varies inversely as the square of its distance from the sun.

Solution. We assume that $T^2 = kr^3$, and that F, the gravitational force between the planet and the sun, provides the required centripetal force and is given by $F = mv^2/r$, where r is the radius of the orbit and v is the speed. $T = 2\pi r/v$ and therefore $T^2 = (2\pi r/v)^2 = kr^3$. If we use this equation and $F = mv^2/r$ to eliminate the velocity, we obtain $F = 4\pi^2 m/(kr^2)$, which is the desired inverse square relationship.

It is true that Newton got considerable help from the work of Kepler and Brahe, and also true that the concept of gravity and the idea that gravity might extend out beyond the surface of the earth were not exclusively due to Newton. These facts, however, in no way detract from the greatness of his contribution. Using the calculus, developed independently by himself and the German mathematician, Gottfried Leibnitz, Newton was able to show specifically that if the planets move in paths which are conic sections—i.e., circles, ellipses, parabolas or hyperbolas— the force experienced by the planet must vary inversely as the square of the distance from some center of force through which the line of action of the force always passes. Such a force is called a central force and for the solar system, this center of force is, of course, the sun.

When one thinks about using equation 8.1 to calculate the gravitational force of attraction between two objects, the question soon arises as to how the distance between them should be measured. It turns out that for objects with a spherically symmetrical distribution of mass, one may assume that all of the mass is concentrated at the center. Newton needed the calculus to prove this and it is interesting to note that Newton's uncertainty about this point probably delayed the publication of his *Principia* by about twenty years.

In thinking about the validity of his inverse square law, Newton made a calculation which it is interesting to repeat. Assuming that the moon moves in a circular path about the earth, one should be able to calculate its centripetal acceleration from its period and the radius of its orbit. This centripetal acceleration should be equal to the value of g experienced by the moon if the gravitational force provides the required centripetal force. Since the center of the moon is about 60 earth radii from the center of the earth, one would expect to find that the moon's acceleration is $1/(60)^2 g = (1/3600) \times 9.8 = 0.00272$ m/sec^2, if the gravitational force between two objects varies as $1/r^2$. The moon takes 27.3 days to go around the earth once. Thus, its period is $27.3 \times 24 \times 60 \times 60 = 23.59 \times 10^5$ sec. The average distance from the center of the earth to the center of the moon is about 239,000 miles or 384.5×10^6 meters, from which $v = 2\pi r/T = 1024$ m/sec. The centripetal acceleration of the moon as it moves around the earth is therefore

$$a = v^2/r = (1024)^2/(384.5 \times 10^6) = 0.00273 \text{ m/sec}^2,$$

which is in agreement with the result obtained above, assuming an inverse square variation for the gravitational interaction between the moon and the earth.

We are accustomed to thinking of the gravitational interaction as a rather strong one because it has such an important part in our everyday life. Why then, do we find it such an effort to lift tables, chairs and suitcases in competition with the earth's gravitational attraction, and yet observe no detectable gravitational attraction between these objects. The answer is, of course, that G, the universal gravitational constant, is extremely small while the mass of the earth is huge (5.98×10^{24} kg). For example, a 120 lb (54.75 kg) blonde cheerleader whose center of mass is 1 meter from that of a 200 lb (91.25 kg) fullback would experience a force of attraction (gravitational) according to equation 8.1 of

$$F = \frac{Gm_1m_2}{r^2} = \frac{6.67 \times 10^{-11} \times 54.75 \times 91.25}{1}$$

$$= 3.33 \times 10^{-7} \text{ newtons,}$$

or about 7.46×10^{-6} lbs.

It is interesting to note that in nature there seem to be four basic types of interaction resulting in four different types of "forces." The strongest of these is the so-called strong interaction which is responsible for the force which holds the nucleus together. If we designate the "strength" of this interaction as 1, the electromagnetic interaction which holds atoms and molecules together and is responsible for the forces between electric charges, has a strength of about 10^{-2}. Next in line is the weak interaction which is responsible for the decay of certain particles, including radioactive nuclei. The weak interaction is indeed weak, having a strength of 10^{-14} on our scale. The gravitational interaction, however, is by far the weakest, being only about 10^{-41} as "strong" as the strong interaction.

8-2. Measurement of G. Despite the very small gravitational forces of attraction between objects of moderate size, the universal gravitational constant G can be measured using the balance shown in Figure 8-2. This torsional balance was conceived by the English scientist, Sir Henry Cavendish, roughly a hundred years

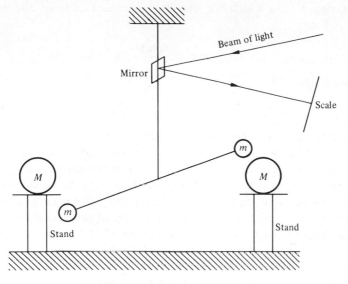

Figure 8-2. The Cavendish Balance.

after Newton proposed his law of gravitation. The principle of operation is quite simple. Two small spheres of mass m, located at the ends of a light rod, are suspended at the center of the rod by a fine fiber. The fiber is extremely sensitive and the slightest twist will cause the whole assembly to rotate. Larger spheres of mass M, are placed as indicated in the figure so as to produce a torque about the suspending fiber. Although small, the resulting twist of the balance can be observed. Often a mirror is mounted on the fiber or on the rod and a narrow beam of light reflected from it onto a conveniently located scale. This arrangement is often called an optical lever. A very slight rotation of the mirror results in an appreciable movement of the spot on the scale if the distance between mirror and scale is fairly large. The balance must be calibrated, since it is obviously essential to know the deflection on the scale produced by a known force. If M, m, and the distance between their centers are known, equation 8.1 can be used to obtain G, if the value of F is provided by the Cavendish experiment.

Clearly, there must be a relationship between g, the acceleration due to gravity at the surface of the earth, and G. This is easily obtained if we equate the force of gravity on an object of mass m, i.e., mg, to the expression for the same quantity using equation 8.1, in which we use the average radius of the earth for r. Thus:

$$mg = \frac{GmM}{r_e^2}$$

from which,

$$G = \frac{gr_e^2}{M} \tag{8.2}$$

where r_e is the mean radius of the earth, and M is the mass of the earth. It is interesting to note that the Cavendish experiment also amounts to a determination of the mass of the earth, since g and the radius of the earth are well known. Knowing the mass and the volume, one can compute the average density of the earth, and the result is about 5500 kg/m^3. Measurements of the average density of the earth's crust give a value of about 2700 kg/m^3, indicating that the interior portions of the earth must have a density even greater than 5500 kg/m^3. Thus, the Cavendish experiment provided not only a measure of G and the mass of the earth but also some geologic information which could be obtained in no other way until the invention of the seismograph some years later.

The reason that the acceleration due to gravity does not change very rapidly with altitude at the surface of the earth is provided by equation 8.2. The value of r to be used is measured from the center of the earth. The radius of the earth is about 4000 miles or 6.36×10^6 meters. Thus, at an altitude of 100,000 meters or 328,000 ft, or about 61 miles, the acceleration due to gravity is reduced to only 9.5 m/sec^2. By human standards, 100,000 m is a substantial change in altitude, but it corresponds to only a 1.6% change in altitude as measured from the center of the earth.

8-3. Gravitational Potential Energy and Earth Satellites. Near the surface of the earth, the gravitational force is essentially constant, and the change in gravitational potential energy of an object of mass m, raised through a vertical height h, is simply *mgh*.

As we pointed out above, this is a good approximation for many miles above the surface of the earth, but as soon as h becomes an appreciable fraction of the earth's radius we must worry about the inverse square dependence of the gravitational force. For example, when $h = 0.2r$, $g = \dfrac{9.8}{(1.2)^2} = 6.8$ m/sec, and for $h = r$ and $h = 2r$, the acceleration due to gravity becomes $\frac{g}{4}$ and $\frac{g}{9}$ respectively, since the distance from the center of the earth for the last two values of h are 2 and 3 earth radii, respectively.

Let us calculate the difference in gravitational potential energy for an object which is taken from point A at the surface of the earth in Figure 8-3 to point B, which is assumed to be many earth

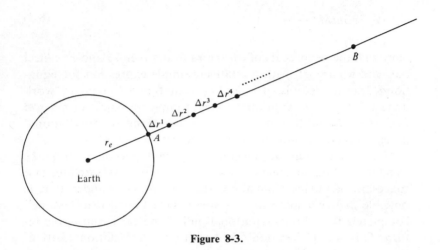

Figure 8-3.

radii from A. Certainly we must take account of the fact that the gravitational force exerted by the earth on this object is not constant. We do this by breaking the calculation up into many steps. We imagine that the object, of mass m, is moved along the line between A and B a very short distance Δr, grossly exaggerated in Figure 8-3. For this short distance, $F = G \dfrac{mM}{r^2}$ is very nearly constant. We must calculate the work done in moving this distance which is $r_1 - r_A$. The work of course equals the force times the distance. In the above expression for the force we replace r^2 with $r_1 r_A$ to obtain an average value for r^2 which is not constant over this small displacement. Our expression for the work becomes:

$$W = F \times r = GmM \frac{r_1 - r_A}{r_1 r_A}$$

where M is the mass of the earth.

The following expression results when we calculate the total work done in a series of many such small displacements Δr.

$$W = GmM \left(\frac{r_1 - r_A}{r_1 r_A} + \frac{r_2 - r_1}{r_2 r_1} + \frac{r_3 - r_2}{r_3 r_2} + \cdots \frac{r_B - r_n}{r_B r_n} \right)$$

$$W = GmM \left(\frac{1}{r_A} - \frac{1}{r_1} + \frac{1}{r_1} - \frac{1}{r_2} + \frac{1}{r_3} + \cdots - \frac{1}{r_B} \right)$$

$$W = GmM \left(\frac{1}{r_A} - \frac{1}{r_B} \right). \tag{8.3}$$

Note that the reciprocals of all the radii between r_A and r_B cancel out, and we are left with a relatively simple expression for equation 8.3, in terms only of r_A and r_B. Equation 8.3 gives the work done against the gravitational force to move an object of mass m from A to B, and therefore also gives the change in gravitational potential energy in moving between these two points.

The calculation in the preceding paragraph permits us to make some interesting but simple calculations about earth satellites and spaceships launched from the earth. One might wonder if it is possible to give a spaceship a velocity such that it can "escape" completely from the gravitational pull of the earth and never return. This might seem doubtful because the gravitational attraction of the earth extends indefinitely, although it becomes extremely weak at large distances. Equation 8.3 permits the calculation of this escape velocity. If we set $r_B = \infty$, $1/r_B$ becomes zero and we have the amount of energy required to move an object an infinite distance from the surface of the earth. To calculate the "escape" velocity we equate the kinetic energy of the spaceship to the required change in potential energy:

$$\tfrac{1}{2} mv^2 = \frac{GmM}{r_e}$$

where r_e is the radius of the earth. Since $G = gr_e^2/M$ we get $v^2 = 2gr_e$. Using the radius of the earth as 6.37×10^6 m, we obtain:

$$v = \sqrt{2 \times 9.8 \times 6.37 \times 10^6} = 11.2 \times 10^3 \, \text{m/sec}$$

$$= 6.94 \text{ miles/sec}$$

or about 25,000 mph.

Example 8-2.   A projectile is shot vertically into the air with a velocity of 5 kilometers per second. How high above the surface of the earth does it rise?

Solution.   We equate the original kinetic energy to the change in potential energy experienced in going from the surface of the earth (at $r = r_e$) to the final radius r, where r is measured from the center of the earth.

$$\tfrac{1}{2}mv^2 = GmM \left(\frac{1}{r_e} - \frac{1}{r} \right)$$

$$r = \frac{2GMr_e}{2GM - v^2 r_e}$$

$$m_e = 5.98 \times 10^{24} \text{ kg}$$

$$r = \frac{2 \times 6.67 \times 10^{-11} \times 5.98 \times 10^{24} \times 6.37 \times 10^6}{2 \times 6.67 \times 10^{-11} \times 5.98 \times 10^{24} - 25 \times 10^6 \times 6.37 \times 10^6}$$

$$r = 7.96 \times 10^6 \text{ m}.$$

Since the radius of the earth is 6.37×10^6 meters, the projectile rises to a height of 1.59×10^6 meters above the surface of the earth. It is interesting to note that this height is independent of the mass of the projectile.

Example 8-3.   Calculate the velocity required of a satellite if it is to move in a circular path around the earth near the surface of the earth. Also calculate its period.

Solution.   The gravitational force on the satellite, mg, must provide the required centripetal force. Thus, $mg = mv^2/r_e$, and $v = \sqrt{r_e g}$. It is interesting that this is equal to the escape velocity divided by the $\sqrt{2}$, or about 17,700 mph. Assuming the radius of the earth to be about 4000 miles, the period of such a satellite would be $T = 2\pi r_e/v = 6.28 \times 4000/17,700 = 1.42$ hrs $= 85.2$ min.

Example 8-4.   Determine the height above the surface of the earth which a satellite must have if it is to move in a circular orbit above the equator with a period equal to that of the earth on its axis; i.e., it remains above the same point on the earth. The mass of the earth $= 5.98 \times 10^{24}$ kg.

Solution.   We set the gravitational force of attraction at radius r (measured from the center of the earth) equal to the centripetal force.

$$\frac{mv^2}{r} = \frac{GmM}{r^2}.$$

Here M is the mass of the earth, while m is the mass of the satellite. We also know that $v = 2\pi r/T$, where T is the period which must equal the period of the point on the earth immediately below it, or 24 hours. Combining these two equations and eliminating v we obtain:

$$r^3 = \frac{GMT^2}{4\pi^2} = \frac{6.67 \times 10^{-11} \times 5.98 \times 10^{24} \times (24 \times 60 \times 60)^2}{4 \times (3.14)^2}$$

$$r^3 = 75.5 \times 10^{21}$$

$$r = 4.23 \times 10^7 \text{ meters} = 2.63 \times 10^4 \text{ mi} = 26,300 \text{ miles}.$$

Note that this is Kepler's third law, and that we can evaluate the constant of proportionality, $GM/4\pi^2 = 1.011 \times 10^{13}$ m^3/sec^2 for a circular orbit. Using the result of Example 8-2 we can use Kepler's third law and obtain the same result somewhat more easily, also using the fact that the radius of the earth is about 4000 miles.

$$\frac{T_1^2}{T_2^2} = \frac{Kr_e^3}{Kr^3} = \frac{(4000)^3}{r^3} = \frac{(1.42 \text{ hr})^2}{(24 \text{ hr})^2}$$

$$r^3 = (4000)^3 \times \left(\frac{24}{1.42}\right)^2 = 18.28 \times 10^{12}$$

$$r = 2.63 \times 10^4 = 26,300 \text{ miles}$$

which puts the satellite about 22,300 miles above the surface of the earth.

Example 8-5. Calculate the acceleration due to gravity at the surface of the planet Jupiter, knowing that its mass is 1.9×10^{27} kg, and its mean radius 6.98×10^7 m.

Solution. We use equation 8.2 and obtain

$$g = \frac{GM}{r_e^2},$$

which gives:

$$g = 6.67 \times 10^{-11} \times 1.9 \times \frac{10^{27}}{(6.98 \times 10^7)^2}$$

$$g = 26.0 \text{ m/sec}^2.$$

Problems

Chapter 2.

1. How far will a car travel in 2 hours if it moves at the constant speed of 40 ft/sec?

    Ans.   54.55 miles.

2. A car travels at the constant speed of 35 mph for 30 miles, 45 mph for the next 30 miles, and 55 mph for the final 30 miles. What was the average velocity for the trip?

    Ans.   43.5 mph.

3. A racing car traveling at a speed of 10 m/sec increases its speed at constant acceleration until its speed is 50 m/sec. If this process takes 20 sec, what was the acceleration?

    Ans.   2 m/sec^2.

4. A driver traveling at a speed of 44 ft/sec suddenly realizes that he must stop for a railroad crossing 150 ft away. He knows that the maximum acceleration possible for his car is 8 ft/sec^2. Can he stop in time and how far in front of or beyond the tracks does he stop?

    Ans.   Yes, 29 ft in front.

5. Calculate the acceleration of a car which goes through a toll booth at 2 ft/sec and then increases its speed at constant acceleration until it has gone 66 ft and has a velocity of 20 ft/sec.

    Ans.   3 ft/sec^2.

6. A boy on a bridge above a river drops a stone which takes 4 seconds to hit the water. How high above the water was the bridge?

    Ans.   256 ft.

7. With what speed must an object be thrown vertically so that it reaches a maximum height of 400 ft?

    Ans.   160 ft/sec.

8. Two cars start from rest at the same time and move in a straight line in the same direction. One has a constant ac-

celeration of 3 ft/sec^2 and the other has a constant acceleration of 4 ft/sec^2. What is the distance between them after
30 seconds?

Ans.   450 ft.

9. Calculate the number of seconds by which astronomers to-
    day would be in error if they estimated the date and hour
    of an eclipse occurring 2000 years ago and neglected to
    account for the fact that the period of rotation of the earth
    on its axis has been steadily increasing at the rate of 0.001
    seconds per century.

Ans.   7300 sec.

10. A motorcycle policeman sees a speeder approaching and
    starts after him just as the speeder passes him. Assuming
    that the policeman starts from rest and moves with a con-
    stant acceleration of 2 ft/sec^2, what was the speed of the
    car if 100 seconds are required to overtake it?

Ans.   100 ft/sec.

11. A ball is thrown straight down from a building with an
    initial velocity of 10 ft/sec. If the building is 450 ft high,
    how long was the ball in the air?

Ans.   5 sec.

Chapter 3.

1. Calculate the magnitude and direction of the vector sum of
    the four vectors shown in Figure P-1.

Ans.   100 lbs at angle of 53° with *x* axis.

2. Measurements from the ground indicate that an airplane is
    heading due north, with respect to the ground, with a speed
    of 100 mph. If the wind is directed 53° south of east and has
    a speed of 40 mph, what is the magnitude and direction of
    the velocity of the plane with respect to the air?

Ans.   134.2 mph, 10.3° west of north.

3. A projectile is launched at an angle of 37° above a level,
    horizontal surface. Calculate how high it rises, how long it
    is in the air and its horizontal range if its initial velocity
    is 80 ft/sec?

Ans.   36 ft, 3.0 sec, 192 ft.

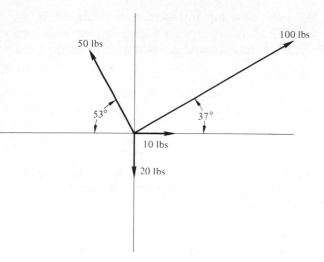

Figure P-1.

4. As shown in Figure P-2, a projectile with initial velocity 240 ft/sec is launched at an angle of 53° with the horizontal. At what distance d does it land from the edge of a level plateau located 64 ft above the original surface? (Sin 53° = 0.8, cos 53° = 0.6.)

Ans. 679 ft.

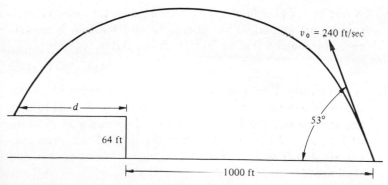

Figure P-2.

5. A boat moves at a speed of 5 ft/sec with respect to the water. It heads directly across a 100 ft wide river in which the current is a steady 3 ft/sec. How long does it take to cross

the river, how far downstream is it carried in the process (relative to where it would have landed with zero current) and what distance was traveled in crossing the stream?

Ans. 20 sec, 60 ft, 117 ft.

6. At what angle must a projectile be launched if its initial velocity is 500 ft/sec and its maximum height is to be 1953 ft?

Ans. 45°.

7. A car, traveling at the constant speed of 44 ft/sec, makes a 90° turn in 5 sec. Calculate the average acceleration experienced by the car.

Ans. 12.44 ft/sec.

8. A train traveling with a speed of 88 ft/sec, goes into a curve of radius 323 ft. A weight, suspended on the end of a string, would normally hang in the vertical direction. Under the above conditions, what angle does it make with the vertical?

Ans. 37°.

9. A roller coaster car goes down a steep hill and into a curved loop at the bottom. If the radius of curvature of the track is 100 ft, what is the acceleration of the car at the bottom of this loop if its velocity is 50 ft/sec?

Ans. 25 ft/sec^2.

10. A fast merry-go-round turns with an angular velocity of 2 radians per second. What horizontal acceleration is experienced by someone riding 20 ft from the axis of rotation?

Ans. 80 ft/sec^2.

11. A gun mounted on a moving flat car is fired vertically when the car is moving at 80 ft/sec. As the shell is fired, the brakes are applied and the flat car starts to slow down at the rate of 2 ft/sec^2. When the shell hits the ground the shell and car are 400 ft apart. With what vertical velocity was the shell fired?

Ans. 320 ft/sec.

Chapter 4.

1. A 96 lb box rests on a horizontal surface. The coefficient of friction between the box and the surface is 0.2. What

horizontal force is required to give the box an acceleration
of 5 ft/sec²?

Ans.   34.2 lbs.

2. A block slides down an inclined plane which makes an angle
of 37° with the horizontal. If the coefficient of friction between the block and the plane is 0.3, calculate the acceleration of the block.

Ans.   0.36g.

3. With what acceleration must an elevator move upward in
order that the "apparent weight" of a passenger be increased
by 50%?

Ans. $a = 0.5g$.

4. What constant horizontal force is required to bring a 1600
lb car moving at 44 ft/sec to a stop in 2 sec?

Ans.   1100 lbs.

5. As shown in Figure P-3, a 10 kg block and a 20 kg block
are connected by a cord of negligible mass and slide down an

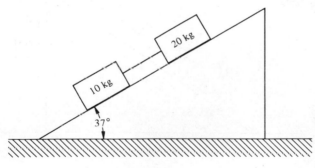

Figure P-3.

inclined plane making an angle of 37° with the horizontal.
If the coefficient of friction for the 10 kg block is 0.2 and
that for the 20 kg block is 0.4, calculate their acceleration
and the tension in the cord as they slide.

Ans.   3.27/m/sec² and 10.42 newtons.

6. Calculate the acceleration of the weights on the Atwood
machine shown in Figure P-4 and calculate the tension in
the cord.

Ans.   10.67 ft/sec² and 21.33 lbs.

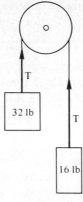

Figure P-4.

7. Figure P-5 shows a 20 kg block being pulled up a 53° inclined plane by an 18 kg mass suspended from a frictionless

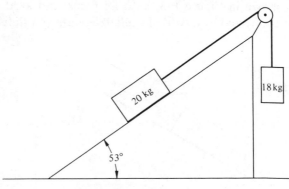

Figure P-5.

pulley. If the coefficient of friction between the 20 kg block and the plane is 0.1, calculate the acceleration of the two blocks and the tension in the cable.

Ans.   0.206 m/sec and 172.7 newtons.

8. A curve on a race track is to be designed for cars traveling at 120 mph (176 ft/sec). If the radius of this curve is 966 ft, what is the proper angle of banking?

Ans.   45°.

9. Consider a conical pendulum as shown in Figure 4-8 (page 52). Calculate the tension in the cord and the angle θ if the length is 1 meter, the mass is 2 kg and ω = 4 radians per second.

Ans. 32 newtons and 52.23°.

10. An automobile travels at 96 ft/sec. If the coefficient of friction between the tires and road is 0.5, calculate the shortest distance in which it can be stopped.

Ans. 288 ft.

11. Figure P-6 shows a 64 and a 32 lb block suspended on frictionless planes. Calculate the tension in the cable and the acceleration of the blocks. Which way does the system move?

Ans. 27.74 lbs and 2.13 ft/sec² to left.

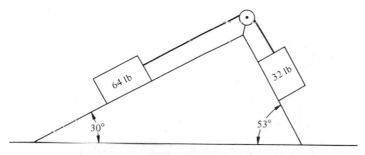

Figure P-6.

12. Consider a projectile launched at some angle θ with the horizontal. What is its acceleration immediately after being launched, at its highest point and shortly before it hits the ground?

Ans. 32 ft/sec² in all cases.

Chapter 5.

1. 500 lbs is suspended as shown in Figure P-7. Calculate the tension in each of the three supporting cables *a*, *b* and *c*.

Ans. 366 lbs, 448 lbs, and 500 lbs.

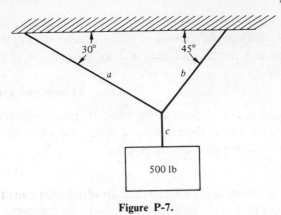

Figure P-7.

2. Calculate the tension in the cable supporting the boom in Figure P-8. The boom is uniform and has a mass of 100 lbs.

Ans. 2188 lbs.

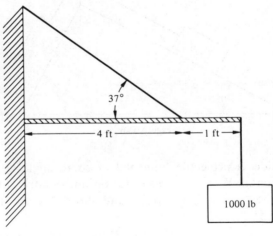

Figure P-8.

3. Calculate the x and y components of the force exerted by the wall on the boom of Figure P-8.

Ans. $R_x = 1750$, $R_y = 213$ lbs (down).

4. Find the tension in the cable of Figure P-9. Neglect the weight of the wooden boom.

Ans. 2000 lbs.

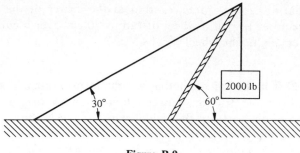

Figure P-9.

5. A 180 lb man hangs from the middle of a rope as illustrated in Figure 5-3 (page 60). The rope is tightly stretched and only makes an angle of 3° with the horizontal. Calculate the tension in the rope.

Ans. 1720 lbs.

6. Two men support a 10 ft board horizontally. Man *A* is on the left and man *B* on the right. Weights of 50 lbs, 60 lbs, and 20 lbs are located 2 ft, 4 ft, and 6 ft from man *A*. What weight must be placed 8 ft from man *A* if *A* and *B* are to carry equal loads?

Ans. 63.3 lbs.

7. A man wishes to use a 6 ft board as a lever to raise a shed from its foundation as indicated in Figure P-10. If the man

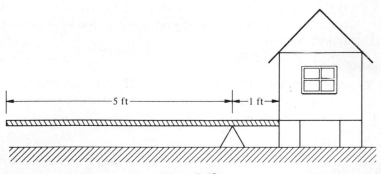

Figure P-10.

weighs 200 lbs and puts all his weight at the far end of the board, what force can be exert on the shed?

Ans. 1000 lbs.

8. What horizontal force, exerted at the center of the wheel, is required to pull a wheel of radius 30 cm over a curbstone 10 cm high? The wheel weighs 5 lbs.

                                                        Ans.   5.59 lbs.

9. A 20 ft ladder of negligible weight leans against a smooth wall (i.e., no friction between ladder and wall). A 200 lb man climbs 15 ft up the ladder which touches the wall 16 ft above the ground. What horizontal force is needed at the foot of the ladder to keep it from slipping?

                                                        Ans.   112.5 lbs.

Chapter 6.

1. An object is dropped from a height of 81 feet. Using the energy principle, calculate its velocity when it reaches the ground.

                                                        Ans.   72 ft/sec.

2. A 4 lb ball is placed on the end of a coil spring of force constant 80 lbs/ft. The spring is compressed 2 feet and then released so that it shoots the ball vertically into the air. How high does the ball rise above its original position?

                                                        Ans.   40 ft.

3. A railroad car coasts down a long hill and then up a smaller one onto a level surface where it has a speed of 32 ft/sec. If the car started 200 feet above the lowest point on the track, how far above this lowest point is the level surface? Ignore friction.

                                                        Ans.   184 ft.

4. A car weighing 4400 lbs climbs a hill in 10 seconds. If the net vertical rise is 200 ft, what is the average horsepower delivered by the engine?

                                                        Ans.   160 hp.

5. A 50 gram bullet moving at 400 m/sec penetrates a block of wood to a depth of 10 cm. What average force does the bullet exert on the block?

                                                        Ans. 4×10^4 newtons.

6. A water pump motor has a horsepower rating of 11.3 hp. If it is to pump water from a well at the rate of 1 ft³/sec (water has a weight density of 62.4 lbs/ft³), what is the maximum depth the well may have? Neglect friction in your calculations.

Ans.   100 ft.

7. A block starting from rest slides a distance of 5 meters down an inclined plane which makes an angle of 37° with the horizontal. The coefficient of friction between block and plane is 0.2. What is the velocity of the block after sliding 5 meters? What would the velocity be if the coefficient of friction were negligible?

Ans.   6.57 m/sec and 7.67 m/sec.

8. A sphere rolls down an inclined plane through a vertical drop of 14 ft. The moment of inertia of a sphere is $\frac{2}{5}mr^2$. What is the velocity of the sphere at the bottom of the incline?

Ans. $v = \sqrt{10gh/7} = 25.3$ m/sec.

9. A circular disc has a moment of inertia of $\frac{1}{2}mr^2$. Calculate the kinetic energy of a circular disc having a mass of 24 kg, a radius of 50 cm, and an angular velocity of 2 radians/sec.

Ans.   6 joules.

10. A boy pushes his friend on a bicycle with a constant force of 6 lbs so that his acceleration is 2 ft/sec². If the bicycle starts from rest, what power does the boy deliver to the bicycle 10 seconds after he starts pushing?

Ans.   120 ft lbs/sec or 0.218 hp.

11. A weight is suspended from a cord 1 meter long and is made to swing back and forth as a pendulum. If the angle between the cord and the vertical direction is 60° at the highest point in the swing, what is the velocity of the weight as it passes through its lowest point, what is its acceleration at this point and what is its acceleration when it is at its highest point?

Ans.   3.13 m/sec, g upward, 8.49 m/sec².

Chapter 7.

1. A 32 lb wagon rolls along in a straight line, with a velocity of 12 ft/sec. A 64 lb boy hanging from a tree limb drops into the wagon. The boy and wagon continue together in the direction in which the wagon was originally moving. Calculate their speed.

   Ans.   4 ft/sec.

2. A rifle of mass 4 kg fires a bullet with a velocity of 640 m/sec. If the mass of the bullet is 50 grams, what is the initial recoil velocity of the gun?

   Ans.   8 m/sec.

3. A soldier, tired of skating, fires his machine gun parallel to the surface of a frozen lake. If the bullets each have a mass of 40 grams and leave the gun at a speed of 500 m/sec at the rate of 10/sec, what average force does the soldier experience and, neglecting friction, what is his acceleration if his total mass is 100 kg?

   Ans.   200 newtons and 2 m/sec².

4. A 6 gram bullet having a speed of 500 m/sec is fired horizontally into a 1.5 kg block on a level surface. The bullet is stopped in the block. How far across the surface does the block slide if it experiences a constant frictional force of 1 newton?

   Ans.   3 m.

5. A boy weighing 64 lbs approaches his 160 lb father on ice skates. If father and son have equal and opposite velocities of 10 ft/sec what is their final velocity if the father picks up the boy as they meet and they move off together?

   Ans.   4.28 m/sec in direction of father's original velocity.

6. A New York City fireboat is pumping water from two hoses on its bow. The water leaves the hoses horizontally at a speed of 200 feet/sec. If each hose pumps 2 cubic feet of sea water (64 lbs/ft³) per second, what thrust must the propellers provide to keep the boat at rest?

   Ans.   1600 lbs.

7. A rifle bullet having a mass of 0.08 kg is fired into a pendulum "bob" having a mass of 12 kg. If the "bob" rises a vertical distance of 10 cm after the collision, calculate the velocity of the bullet.

Ans. 211.4 m/sec.

8. A 50 gram rifle bullet moving horizontally strikes a 2 kg block resting on a level table. The bullet passes through the block and measurements indicate that it emerges with a velocity of 100 m/sec in the same direction. Immediately after the collision, the block is found to have a velocity of 10 m/sec in the same direction. Calculate the original velocity of the bullet.

Ans. 500 m/sec.

9. A hunter suddenly observes that a large duck having a mass of 2 kg is falling vertically directly above him at the rate of 10 m/sec. Being proud of his marksmanship and his physics he rapidly fires a certain number of shots from his rifle vertically upward, rather than step aside. The velocity of the bird is reduced to zero directly above the hunter and the bird is easily caught. How many shots did the hunter have to fire if one assumes that the bullets remained in the bird, had a mass of 50 grams and a velocity of 200 m/sec?

Ans. 2 shots.

10. A rod of negligible mass and length 2 meters has small 2 kg masses mounted on each end. Calculate the moment of inertia of this rod about an axis perpendicular to the rod and through its center and its angular momentum if the rod rotates about this axis with an angular velocity of 10 radians per second.

Ans. 4 kg m^2 and 40 kg m^2/sec.

11. A man with weights in his hands and his hands pressed close to his body stands on a turntable which rotates with an angular velocity of 2 radians per second. His total moment of inertia in this position is 4 kg m^2. He then extends his arms and in this position his total moment of inertia becomes 6 kg m^2. What is his new angular velocity?

Ans. 1.33 rad/sec.

Chapter 8.

1. Calculate the force of attraction in newtons between two spherical cannon balls each of which has a mass of 50 kg and whose centers are separated by a distance of 50 cm.

   Ans. 6.67×10^{-7} newtons.

2. The mass of the earth is 5.98×10^{24} kg and its radius is 6380 km. If the planet Mars has a mass equal to 0.106 times that of the earth and a radius of 3430 kilometers, and the planet Jupiter has a mass equal to 314.5 times that of the earth and a radius of 71,800 kilometers, calculate the acceleration due to gravity on the surface of each planet.

   Ans.   3.59 m/sec² on Mars and 24.33 m/sec² on Jupiter.

3. Using the information provided in Problem 2, calculate the periods of satellites moving in circular orbits about Mars and Jupiter slightly above their surfaces.

   Ans.   102.2 min for Mars and 179.8 min for Jupiter.

4. If the period of the earth about the sun is 365 days and the average distance from the earth to the sun is 92.9 million miles, calculate the period of Jupiter about the sun if its mean distance from the sun is 483.4 million miles.

   Ans.   11.9 years.

5. Knowing that the mass of the moon is 0.0123 times that of the earth, calculate the fractional distance from the earth along the line connecting the earth and the moon that one must go before the gravitational attraction of the moon equals that of the earth.

   Ans.   0.9 of distance.

6. Calculate the mass of the earth, given the information that the period of the moon about the earth is 27.3 days and the radius of the moon's orbit about the earth is 38×10^4 kilometers.

   Ans. 5.80×10^{24} kg.

7. Calculate the acceleration due to gravity on the moon given that its radius is one fourth that of the earth and its mass is about 0.0123 that of the earth.

   Ans.   6.30 ft/sec².

8. Using the information provided in Problem 2, calculate the escape velocities from the planets Mars and Jupiter.

    Ans. 4.96 × 10^3 m/sec for Mars and
    5.91 × 10^4 m/sec for Jupiter.

9. The period of the moon about the earth is 27.3 days. If the radius of its orbit is 23.9 × 10^4 miles, at what distance from the earth would another moon or satellite have a period of 4.55 days?

    Ans. 7.24 × 10^4 miles.

10. Show that the acceleration due to gravity at any height h above the surface of the earth is given by:

    $$g_h = \frac{g}{\left(1 + \frac{h}{r}\right)^2}$$

    where r is the radius of the earth.

11. Calculate the velocity which an object must have in order to rise vertically a distance above the surface of the earth of 3 earth radii. The radius of the earth is 6.37 × 10^6 meters and its mass is 5.98 × 10^{24} kg. Neglect air resistance.

    Ans. 9.69 × 10^3 m/sec.

Index